D0530077

BrightRED Revision

Higher CHEMISTRY

Archie Gibb and David Hawley

First published in 2008 by:

Bright Red Publishing Ltd
6 Stafford Street
Edinburgh
EH3 7AU

Copyright © Bright Red Publishing Ltd 2008

All rights reserved. No part of this publication may be reproduced, stored in a retrieval system, or transmitted in any form or by any means, electronic, mechanical, photocopying, recording or otherwise, without prior permission in writing from the publisher.

The rights of Archie Gibb and David Hawley to be identified as the authors of this work have been asserted by them in accordance with sections 77 and 78 of the Copyright, Designs and Patents Act 1988.

A CIP record for this book is available from the British Library

ISBN 978-1-906736-03-3

With thanks to Ken Vail Graphic Design, Cambridge (layout) and Ailsa Morrison (copy-edit)

Cover design by Caleb Rutherford – eidetic

Illustrations by Beehive Illustration (Mark Turner), Ken Vail Graphic Design.

Acknowledgements

Every effort has been made to seek all copyright holders. If any have been overlooked then Bright Red Publishing will be delighted to make the necessary arrangements.

The publishers would like to thank the following for the permission to reproduce the following images:
© Royal Mail (p67)

Bright Red Publishing would also like to thank the Scottish Qualifications Authority for use of Past Exam Questions. Answers do not emanate from the SQA.

Cover image © Caleb Rutherford

Printed and bound in Scotland by Scotprint.

CONTENTS

HIGHER CHEMISTRY

COURSE STRUCTURE

The Higher Chemistry course is divided into three units:

- Unit 1: Energy Matters
- Unit 2: The World of Carbon
- Unit 3: Chemical Reactions

ASSESSMENT

There are two types of assessment – **internal** and **external**.

For each of the three units, the **internal assessment** consists of a NAB test. You have to gain at least 18 marks out of a possible 30 marks to pass.

Practical Abilities are also assessed internally. This consists of writing a report to a satisfactory standard on **one** of the PPAs from Unit 1.

The **external assessment** consists of an examination paper of duration 2 hours and 30 minutes and with a total allocation of 100 marks.

The exam paper is divided into two sections:

- **Section A** which is worth **40 marks** is made up of 40 multiple-choice questions.

- **Section B** which is worth **60 marks** contains questions which require written answers. In this section of the paper, approximately 6 marks are allocated to questions based on any of the nine PPAs of the course.

Of the 100 marks in the paper, approximately 60 marks are allocated to 'Knowledge and Understanding' (KU) questions and the remainder to 'Problem Solving' (PS) questions.

The course award is graded A, B, C or D depending on how well you do in the external examination. In order to gain the course award, you must also pass the three NABs (one for each unit) and complete a PPA report up to the standard required by the SQA.

AIM AND STRUCTURE OF THIS BOOK

The aim of this Revision book is to help you achieve success in the final exam by providing you with a concise and engaging coverage of the syllabus content.

The book is divided into the three units of the course and within each unit there is a double-page spread on each of the sub-sections.

Each double-page spread:

- covers the content of the sub-section in a logical and digestible manner and will allow you to get a good understanding of the key ideas and concepts.

- contains **'Don't Forget'** sections. These flag up important pieces of knowledge that you need to remember and important things that you must be able to do.

- contains a **'Let's think about this'** section which takes different forms. Some have questions designed to test your knowledge and understanding of the content. Others are designed to extend your knowledge of the subject and provide additional interest. Answers are also provided, either immediately after the questions or on p108.

At the end of each unit there is a section devoted to the three PPAs of that unit. It covers the 'Aim', 'Procedure', 'Results' and 'Conclusion' for each PPA. It also includes a section labelled 'Evaluation' in which important aspects of the PPAs are highlighted.

References have been included throughout the book to the Higher Chemistry Data Booklet. The SQA has a downloadable PDF on their website, go to:
http://www.sqa.org.uk/files_ccc/NQChemistryDataBooklet_H_AdvH.pdf

REACTION RATES 1

FOLLOWING THE COURSE OF A REACTION

Measuring the rate of a reaction

In a chemical reaction reactants change into products. As a chemical reaction proceeds, the reactants are being used and the products are being formed.

Some chemical reactions such as explosions are very fast, whereas corrosion of iron is very slow. Most chemical reactions take place at rates in between these extremes.

The rate of a reaction can be expressed in terms of changes in concentration or mass or volume of the reactants or products per unit time. For example, consider the reaction between marble chips (a form of calcium carbonate) and dilute hydrochloric acid:

$$CaCO_3(s) + 2HCl(aq) \rightarrow CaCl_2(aq) + H_2O(l) + CO_2(g)$$

In this reaction carbon dioxide gas is being produced and, using the apparatus shown in the diagrams below, the **mass** of **carbon dioxide 'lost'** can be measured at regular intervals or the **volume** of **carbon dioxide produced** can be measured at regular intervals.

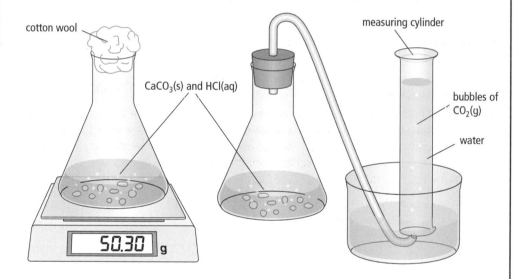

If a gas is produced then the decrease in mass of the apparatus is the same as the mass of gas produced and if the mass is recorded at regular intervals then,

$$\text{Rate of reaction} = \frac{\text{change in mass of the apparatus}}{\text{time taken for the change to occur}}$$

If the time is measured in seconds, then the units for reaction rate will be $g\,s^{-1}$.

If the volume of gas produced is recorded at regular intervals then,

$$\text{Rate of reaction} = \frac{\text{change in volume}}{\text{time taken for the change to occur}}$$

If the time is measured in minutes, then the units for reaction rate will be $cm^3\,min^{-1}$.

Concentration is usually measured in $mol\,l^{-1}$ and if the changes in concentration of a reactant or a product are measured at regular intervals then,

$$\text{Rate of reaction} = \frac{\text{change in concentration of reactant or of a product}}{\text{time taken for the change}}$$

and the units of rate will be $mol\,l^{-1}\,s^{-1}$.

DON'T FORGET

The units of rate depend on whether mass or volume or concentration is being measured at regular time intervals.

contd

FOLLOWING THE COURSE OF A REACTION contd

Calculating the reaction rate from graphs

Examples of graphs of experimental results using the apparatus above are given below.

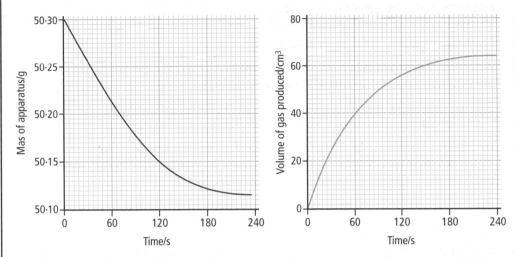

Looking at the graph on the left, it can be seen that at time 0 s, the mass of apparatus is 50·30 g. After 60 s the mass has decreased to 50·21 g. Therefore, the change in mass is 0·09 g over the first 60 seconds.

The average rate of reaction over the first 60 s is $= \frac{0·09}{60} = 0·0015$ g s^{-1}.

From 120 s to 180 s, the mass has changed from 50·15 g to 50·12 g and so the average rate of reaction between 120 s to 180 s is $\frac{0·03}{60} = 0·00050$ g s^{-1}

The slope of the graph gives an indication of the reaction rate. The greater the slope, the faster the reaction. When the reaction is over, the graph line is horizontal and the slope is zero. The slope is at its greatest at the start of the reaction and decreases as the reaction proceeds. The reaction is quickest at the start, because the concentration of the reactants is high, and the reaction slows down as the reactants get used up.

In some reactions, the time, t, taken for an event such as a colour appearing, is measured. In this case the rate of the reaction is taken to be $\frac{1}{t}$.

DON'T FORGET

You must be able to calculate the average reaction rate from a graph.

LET'S THINK ABOUT THIS

Look at the graph above of *volume of gas produced* against *time*. Use it to calculate the average rate of reaction between:

(i) 0 – 60 s

(ii) 60 – 120 s

(iii) 120 – 180 s

(iv) 180 s – 240 s.

For answers, see p108.

REACTION RATES 2

FACTORS AFFECTING THE RATE OF A REACTION

Particle size

Magnesium powder reacts faster than magnesium ribbon. Powdered calcium carbonate reacts faster than calcium carbonate lumps. This is because powder has a smaller particle size than ribbon or lumps. The smaller the particle size then the greater the surface area and the faster the chemical reaction.

Concentration

2 mol l^{-1} hydrochloric acid reacts faster than 0·2 mol l^{-1} hydrochloric acid. The greater the concentration then the more reaction particles in a given volume and the faster the chemical reaction.

Collision theory

Simple collision theory states that, for a chemical reaction to occur, the reacting particles (atoms, molecules or ions) must collide into each other. This explains why increasing the surface area speeds up a chemical reaction. There will be more collisions between the reacting particles and so the reaction will be faster.

The collision theory also explains why increasing the concentration speeds up a chemical reaction. More reacting particles in a given volume mean that more collisions will take place between these reacting particles and again the reaction will be faster.

Temperature

Most chemical reactions take place faster at higher temperatures and many chemical reactions will only take place if they are heated above a certain temperature, or if a spark is used to get the reaction started. If we consider hydrogen gas mixed with oxygen gas in a container, every second there are millions of collisions taking place between the hydrogen molecules and oxygen molecules. However, no chemical reaction will take place until a flame or spark is used to ignite the mixture. Once started, a very fast reaction takes place. This means that the collision theory has to be adapted slightly to take this into account.

Collision theory, temperature and activation energy

A chemical reaction can only occur when the reacting particles collide **with enough kinetic energy**.

Temperature is a measure of the average kinetic energy of the particles of a substance and so increasing the temperature means that the reacting particles now have greater kinetic energy and so collisions between them are likely to be more successful.

The minimum energy required by the colliding particles for a collision to be successful is known as the **activation energy, E_A.**

Not all colliding particles have the same kinetic energy. Some particles may be moving very slowly and some may be moving very quickly. This is shown in the energy distribution diagram opposite.

The particles represented at point A have low kinetic energy. Most particles have energy around the average which is around point B but only those particles with energy greater or equal to the activation energy, E_A will take part in successful

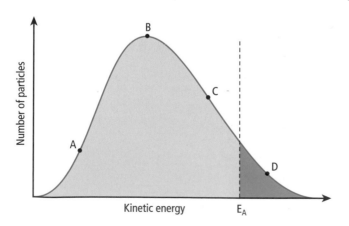

DON'T FORGET

Decreasing particle size and increasing concentration speed up a chemical reaction because there will be more collisions between the reacting particles.

contd

FACTORS AFFECTING THE RATE OF A REACTION contd

collisions. For example, those particles represented at point D have energy greater than E_A but those at points A, B and C do not. The total number of particles with energy greater than the minimum activation energy required is represented by the dark green shaded area to the right of the vertical line representing the activation energy, E_A.

The effect of changing temperature on the kinetic energy is seen in the following diagram. The graph labelled T_1 is the original graph shown above. The graph labelled $T_1 + 10°C$ shows the energy distribution when the temperature has increased by 10°C.

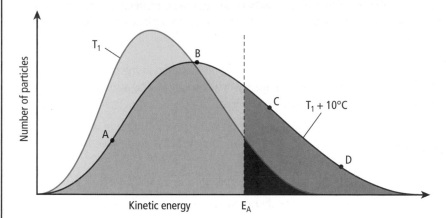

DON'T FORGET

Increasing the temperature means that more particles have energy greater than the activation energy.

At the higher temperature many more particles have energy equal to or greater than the activation energy for the reaction. For example, the particles at point C now have enough energy for successful collisions.

Photochemical reactions

Normally the reacting particles gain more energy by increasing the temperature. However, in photochemical reactions, light provides the energy to the reacting particles so that they have energy equal to or greater than the minimum energy required for successful collisions to take place. An example of a photochemical reaction is an alkane reacting with bromine. In darkness the reaction will not take place, but it can be explosive in light.

LET'S THINK ABOUT THIS

1 When methane gas burns in a Bunsen burner it must first be ignited. Why does it continue to burn once it has been ignited?

2 Why do some reactions such as neutralisation take place instantaneously at room temperature but others such as magnesium burning in oxygen need a flame to get them started?

For answers, see p108.

REACTION RATES 3

FACTORS AFFECTING THE RATE OF A REACTION

Catalysts – homogeneous or heterogeneous?

DON'T FORGET

Heterogeneous catalysts are in a different state compared to the reactants in the reaction being catalysed.

Catalysts **speed up** chemical reactions but at the end of the reaction the catalyst is **unchanged**, in other words it is the same at the end as it was at the beginning of the reaction.

Catalysts can be either **homogeneous** (in the **same state as the reactants**) or **heterogeneous** (in a **different state to the reactants**).

Examples of homogeneous catalysts include cobalt chloride solution in the reaction between potassium sodium tartrate solution and hydrogen peroxide solution, and also enzymes catalysing biochemical reactions.

Examples of heterogeneous catalysts are iron granules in the Haber process in which the reactants are nitrogen and hydrogen gases, and solid manganese(IV) oxide speeding up the decomposition of hydrogen peroxide solution.

Catalyst and the activation energy

All catalysts provide an alternative reaction pathway with lower activation energy. Heterogeneous catalysts provide a surface for the reacting molecules to collide on. The reacting molecules are **adsorbed** on to the surface of the catalyst and form weak bonds with **active sites** on the surface of the catalyst. At the same time the bonds inside these molecules are weakened and so there is more chance of a successful collision when the particles of the other reactant hit them. Less energy is needed and so the activation energy for the reaction has been lowered.

The effect that this has on the number of reacting particles which can have successful collisions is shown below. The red dotted line shows the activation energy for the reaction when no catalyst is present and only those particles in the area shaded orange have enough energy.

The purple dotted line represents the lowered activation energy when a catalyst is used. Now the particles within the area shaded pink, as well as those within the orange area, have energy greater than the activation energy for the reaction and may collide successfully for a reaction to take place.

The blue dotted line represents the lowered activation energy when a more efficient catalyst is used. Now the particles within the area shaded pale blue, as well as those within the orange and pink areas, have energy greater than the activation energy for the reaction.

DON'T FORGET

Catalysts lower the activation energy of the reaction.

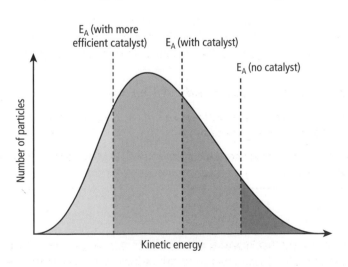

contd

FACTORS AFFECTING THE RATE OF A REACTION contd

Catalysts in Industry

Since catalysts speed up chemical reactions and are unchanged at the end of the reaction, this means that they can be used over and over again. It is important in the chemical industry that the products are made quickly, and using the same catalyst again and again also cuts down on costs. This is important as some catalysts are very expensive.

Some examples of industrial catalysts are shown in the table.

Catalyst	Name of process	Main product formed	Reactants
Iron granules	Haber	Ammonia, NH_3	Nitrogen and hydrogen gases
Nickel	Making margarine	Margarine	Hydrogen and vegetable oils
Platinum	Ostwald	Nitric acid, HNO_3	Ammonia and oxygen
Vanadium(V) oxide	Contact	Sulphuric acid, H_2SO_4	Sulphur dioxide and oxygen

Impurities present with the reactants may block the active sites on the surface of a heterogeneous catalyst and render it useless. This is known as catalyst poisoning. The catalyst then has to be regenerated or even renewed when this happens. This can be a costly process. Therefore, in industry, it is very important that the impurities are removed from the reactants before they enter the reaction chamber which contains the catalyst.

A very common chemical reaction which uses a heterogeneous catalyst is the one which takes place in **catalytic converters** inside exhaust systems of cars. In a petrol engine, poisonous carbon monoxide (made from the incomplete combustion of hydrocarbons) and harmful oxides of nitrogen (made from any nitrogen and oxygen around the spark plug) are produced as unwanted by-products. On the very large surface area inside a catalytic converter, the carbon monoxide and oxides of nitrogen react to form the much less harmful nitrogen and carbon dioxide gases. The catalysts inside a catalytic converter include the expensive metals platinum, rhodium and palladium. Because of this, catalytic converters are very expensive to replace. Nowadays the only petrol sold at service stations is lead-free. One reason that the lead compounds which used to be in petrol had to be removed was that lead poisoned the catalysts inside the catalytic converter.

Enzymes

Enzymes are **biological catalysts**. This means that enzymes catalyse the chemical reactions taking place in the living cells of plants and animals. Enzymes must be present for many of the reactions to take place inside living organisms. Two examples of reactions catalysed by enzymes are:

- The fermentation of glucose into ethanol which is catalysed by zymase;
- Poisonous hydrogen peroxide produced in chemical reactions in our body is broken down into harmless water and oxygen by the enzyme catalase.

Enzymes are also used in many industrial processes such as in the preparation and manufacture of foods, pharmaceuticals and detergents.

LET'S THINK ABOUT THIS

Explain the different way in which a catalyst increases the number of reacting particles with the minimum energy required (the activation energy) compared to how increasing the temperature increases the number of particles with the minimum energy required.

For answer, see p108.

ENTHALPY 1

POTENTIAL ENERGY DIAGRAMS

What is enthalpy?

The **energy content** of a substance is known as its **enthalpy**. Enthalpy is given the symbol, **H**. This is because it used to be taken as a measure of the **heat content** of a substance.

The energy content varies from substance to substance but it is not possible to measure the absolute enthalpy of different substances. However, during a chemical reaction there is an enthalpy change when reactants change into products. It is possible to measure the enthalpy change which takes place in a chemical reaction. This is measured in kJ and when worked out for 1 mole of substance the units are $kJ\,mol^{-1}$.

Enthalpy change is given the symbol, ΔH, and is equal to the difference in enthalpy between the products and reactants, i.e. $\Delta H = H_P - H_R$ where H_P is the enthalpy or energy content of the products and H_R is the enthalpy of the reactants.

Endothermic and exothermic reactions

An **exothermic reaction** is one in which **energy is given out** to the surroundings. The reaction mixture is taken to be part of the surroundings and if a thermometer is placed in the reaction mixture in an exothermic reaction, the **temperature will be seen to rise**. The enthalpy change in an exothermic reaction is equal to the energy released to the surroundings.

Since energy is released in an exothermic reaction, the value of H_P must be lower than the value of H_R.

$\Delta H = H_P - H_R$ and so for an **exothermic** reaction, ΔH must have a **negative** value.

An **endothermic reaction** is one in which **energy is taken in** from the surroundings. If a thermometer is placed in the reaction mixture in an endothermic reaction the **temperature will be seen to fall** as the reaction takes in heat from the reaction mixture itself. The enthalpy change in an endothermic reaction is equal to the energy absorbed from the surroundings.

In an endothermic reaction the value of H_P will be greater than the value of H_R since energy has been taken in as the products are made from the reactants.

$\Delta H = H_P - H_R$ and so for an **endothermic** reaction, ΔH must have a **positive** value.

Most chemical reactions are exothermic.

Examples of exothermic reactions include combustion and neutralisation.

An example of an endothermic reaction is sparking air to get nitrogen and oxygen to combine to form oxides of nitrogen.

Using potential energy diagrams

A potential energy diagram is used to show the energy pathway for a chemical reaction. It shows how the potential energy changes as the reactants change into the products.

The potential energy changes for an exothermic reaction are shown diagrammatically on the right.

The enthalpy change for a chemical reaction can be calculated from the potential energy diagram for that reaction.

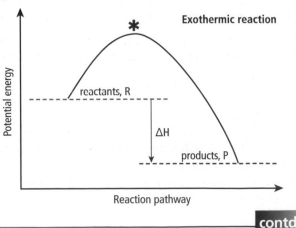

Exothermic reaction

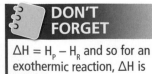

DON'T FORGET

$\Delta H = H_P - H_R$ and so for an exothermic reaction, ΔH is negative.

contd

POTENTIAL ENERGY DIAGRAMS contd

The activated complex

In both exothermic and endothermic reactions there is a barrier which must be overcome before the reactants can change into products. The energy difference between the potential energy of the reactants and the top of this barrier is the activation energy for the forward reaction. The reacting particles must collide into each other with energy equal to or greater than the activation energy for the collisions to be successful. A **successful collision** is when the reacting particles collide with enough energy to form the **activated complex**. The potential energy of the activated complex is at the top of the activation barrier and is represented by ✳ in the potential energy diagrams on page 12 and below.

> The activation energy is the energy required by colliding molecules to form the activated complex.

The value for the activation energy can be calculated from a potential energy diagram. In the potential energy diagram for the endothermic reaction shown below, the activation energy for the forward reaction is shown as E_A.

The same activated complex would be formed in the reverse reaction, that is going from products to reactants. The activation energy for the reverse reaction is shown in the potential energy diagram opposite as $E_A^\#$.

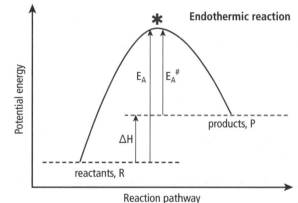

DON'T FORGET

$\Delta H = H_P - H_R$ and so for an endothermic reaction, ΔH is positive.

In all reactions, whether exothermic or endothermic, the activated complex always has a higher potential energy than both the reactants and products. This is because the **activated complex is an unstable arrangement of atoms** compared to both the reactants and products. The potential energy of the activated complex is always at the top of the potential energy barrier.

Consider the reversible reaction, hydrogen and iodine reacting to form hydrogen iodide, and hydrogen iodide breaking down to form hydrogen and iodine. The equation for the reaction showing structural formulae for the reactants and products and the most likely activated complex is:

$$
\begin{array}{ccccc}
H & I & H \cdots I & H-I \\
| & + & | & \rightleftharpoons & \vdots \vdots \vdots & \rightleftharpoons & + \\
H & I & H \cdots I & H-I \\
\end{array}
$$

In the equation above, the solid lines represent covalent bonds and the broken lines represent covalent bonds in the process of being broken or being made.

The activated complex represents an unstable intermediate stage between the reactants and the products, and has only a fleeting existence changing either into the products or back into the reactants.

DON'T FORGET

The same activated complex is formed in the reverse reaction as in the forward reaction.

⚙ LET'S THINK ABOUT THIS

1. Give a reason why some collisions between the reacting particles are not successful in forming the activated complex.
2. Give a reason why, even if the collision is successful in producing the activated complex, the product may not be formed.

For answers, see p108.

ENTHALPY 2

POTENTIAL ENERGY DIAGRAMS AND CATALYSTS

Effect of catalysts

Catalysts lower the activation energy barrier for a reaction so more reacting molecules now have enough energy to collide successfully.

This is shown in the potential energy diagram for an exothermic reaction below.

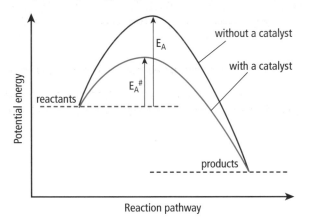

DON'T FORGET

Catalysts lower the activation energy (E_A) of a chemical reaction but do not have any effect on the enthalpy change, ΔH.

E_A is the activation energy of the reaction without the catalyst being present.

$E_A^{\#}$ is the activation energy of the reaction when the catalyst is present.

Notice that the presence of the catalyst does not alter the potential energy or enthalpy of the reactants nor the potential energy or enthalpy of the products. This means that **the enthalpy change for the reaction is not changed** when the catalyst is present.

The same is also true for endothermic reactions as illustrated below:

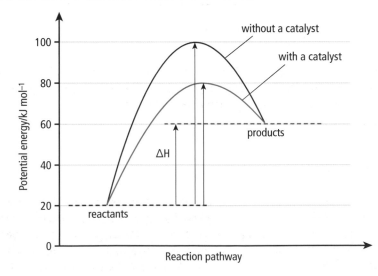

You should be able to calculate from this potential energy diagram that

- $\Delta H = 40\,kJ\,mol^{-1}$
- E_A for the forward uncatalysed reaction $= 80\,kJ\,mol^{-1}$
- E_A for the reverse uncatalysed reaction $= 40\,kJ\,mol^{-1}$
- E_A for the forward catalysed reaction $= 60\,kJ\,mol^{-1}$
- E_A for the reverse catalysed reaction $= 20\,kJ\,mol^{-1}$

contd

POTENTIAL ENERGY DIAGRAMS AND CATALYSTS contd

Alternative reaction pathway

Catalysts lower the activation energy of a chemical reaction by **providing an alternative reaction pathway**. The activated complex in the alternative reaction pathway has a lower potential energy than in the uncatalysed reaction. The catalyst takes part in the reaction but it is unchanged at the end of the reaction.

If we consider the reaction of hydrogen and iodine forming hydrogen iodide, the activation energy is approximately $170 \, kJ \, mol^{-1}$.

The equation for this reaction using structural formula is given on page 13.

However, if platinum is used as a catalyst, then the activation energy drops to about $40 \, kJ \, mol^{-1}$. A possible alternative reaction pathway is:

Note the platinum catalyst takes part in the reaction but at the end it is unchanged.

There are two activated complexes in this alternative reaction pathway, and both are formed quickly and have lower potential energies than the activated complex formed in the uncatalysed reaction.

This is shown below.

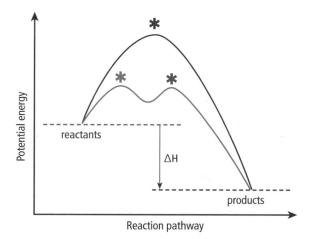

The potential energy of the activated complex in the uncatalysed reaction (✱) is much greater than the potential energies of the activated complexes in the catalysed reactions (✱ and ✱). As before, the enthalpy change is unaffected when a catalyst is used.

LET'S THINK ABOUT THIS

1   In the reaction between potassium sodium tartrate and hydrogen peroxide, the catalyst is cobalt(II) chloride solution. The reaction is slow but as soon as the pink cobalt(II) chloride solution is added the reaction mixture turns green and at this stage the reaction speeds up producing oxygen gas. When the reaction is over, the colour of the solution returns to pink. Explain what is happening here in terms of the Co^{2+} ion acting as a homogeneous catalyst.

2   A chemical reaction has enthalpy change, ΔH, of $-65 \, kJ \, mol^{-1}$ and activation energy, E_A, for the forward reaction of $150 \, kJ \, mol^{-1}$. A catalyst is added which lowers the E_A for the forward reaction to $48 \, kJ \, mol^{-1}$. Calculate
    (i) E_A for the reverse uncatalysed reaction;
    (ii) E_A for the reverse catalysed reaction;
    (iii) ΔH for the reverse reaction.

For answers, see p108.

ENTHALPY 3

ENTHALPY CHANGES

Enthalpy of combustion

The enthalpy of combustion of a substance is the enthalpy change when one mole of the substance burns completely in oxygen.

Substance	Enthalpy of combustion, $\Delta H_c/\text{kJ mol}^{-1}$
Hydrogen	−286
Carbon	−394
Methane	−891
Ethane	−1560
Propane	−2220
Butane	−2877
Methanol	−727
Ethanol	−1367

Enthalpy of combustion values are given in the Data Booklet page 9. The values are always negative since combustion is an exothermic reaction. Units are in kJ mol^{-1} since it is the energy value for one mole of the substance burning completely in oxygen.

Enthalpy of combustion values of some substances are given in the table on the right.

DON'T FORGET

Enthalpy of combustion values are always negative and always quoted for one mole of the substance being burned.

Balanced chemical equations which match enthalpy of combustion values must always contain one mole of the substance being burned. This may mean that these equations contain fractions of moles of other substances, particularly, oxygen. Look at the examples below:

1 mole of hydrogen burning $\qquad$ $H_2(g) + \tfrac{1}{2}O_2(g) \longrightarrow H_2O(l)$
1 mole of ethane burning $\qquad$ $C_2H_6(g) + 3\tfrac{1}{2}O_2(g) \longrightarrow 2CO_2(g) + 3H_2O(l)$
1 mole of methanol burning $\qquad$ $CH_3OH(l) + 1\tfrac{1}{2}O_2(g) \longrightarrow CO_2(g) + 2H_2O(l)$

It is perfectly acceptable to have fractions in balanced chemical equations since it is fractions of moles not fractions of molecules.

Enthalpy of solution

The enthalpy of solution of a substance is the enthalpy change when one mole of the substance dissolves in water.

Enthalpies of solution can be exothermic or endothermic depending on the substance dissolving.

Sodium hydroxide dissolves exothermically in water and so it has a negative enthalpy of solution.

$\qquad$ $Na^+OH^-(s) + aq \longrightarrow Na^+(aq) + OH^-(aq)$ $\qquad$ $\Delta H = -37\,\text{kJ mol}^{-1}$

Ammonium nitrate takes in heat when it dissolves and so has a positive enthalpy of solution.

$\qquad$ $NH_4^+NO_3^-(s) + aq \longrightarrow NH_4^+(aq) + NO_3^-(aq)$ $\qquad$ $\Delta H = +26\,\text{kJ mol}^{-1}$

Enthalpy of neutralisation

The enthalpy of neutralisation of an acid is the enthalpy change when the acid is neutralised to form one mole of water.

When an alkali is added to an acid, the hydrogen ions of the acid react with the hydroxide ions from the alkali.

$H^+(aq) + OH^-(aq) \longrightarrow H_2O(l)$

The other ions present are spectator ions.

Calculating enthalpy values from experimental results

Experiments on enthalpy changes can be carried out in which a known volume or mass of water increases in temperature (exothermic reaction) or decreases in temperature (endothermic reaction).

contd

ENTHALPY CHANGES contd

The heat given out or taken in by water, $E_h = cm\Delta T$, where
c = specific heat capacity of **water** ($4.18\,kJ\,kg^{-1}\,°C^{-1}$ as given in Data Booklet, page 19)
m = mass of **water**, in kg
ΔT = change in temperature of **water**, in °C.

This gives a value in kJ and the enthalpy change is then calculated per mole of substance giving a value in $kJ\,mol^{-1}$.

Calculating enthalpy of combustion from experimental results

Using the apparatus shown, it was found that when 0·34g of ethanol was burned the temperature of 100 cm³ of water rose by 10 °C. Calculate the enthalpy of combustion of ethanol.

Answer

$c = 4.18\ kJ\,kg^{-1}\,°C^{-1}$
Volume of water = 100 cm³ and so the mass of water, m = 100 g = 0·1 kg
$\Delta T = 10\,°C$

Heat taken in by the water, $E_h = cm\Delta T = 4.18 \times 0.1 \times 10 = 4.18\,kJ$

This is the heat given out by 0·34g of ethanol, C_2H_5OH.
One mole of ethanol = 46 g

So the heat given out by one mole of ethanol $= 4.18 \times \dfrac{46}{0.34} = 565.5\,kJ\,mol^{-1}$

Combustion is exothermic and so $\Delta H = -565\,kJ\,mol^{-1}$ from these experimental results. Note that this value is much smaller than the value in the Data Booklet and reasons for this are given on page 35.

Calculating enthalpy of solution from experimental results

In an experiment 4·45 g of solid ammonium nitrate was dissolved in 100 cm³ of water and the temperature of the water dropped from 18·8 °C to 15·4 °C. Use these results to calculate the enthalpy of solution of ammonium nitrate.

Since the temperature fell, the water has lost heat and the process must be endothermic. The enthalpy change will have a positive value.

$\Delta T = 18.8 - 15.4 = 3.4\,°C$
Heat lost by the water, $E_h = cm\Delta T = 4.18 \times 0.1 \times 3.4 = 1.42\,kJ$.

This was using 4·45 g of NH_4NO_3, one mole of which is 80 g.

Therefore the enthalpy of solution for ammonium nitrate is $1.42 \times \dfrac{80}{4.45} = 25.5\,kJ$.

So from these experimental results, the enthalpy of solution of ammonium nitrate = + 25·5 $kJ\,mol^{-1}$.

DON'T FORGET

Learn the formula $E_h = cm\Delta T$ and how to use it.

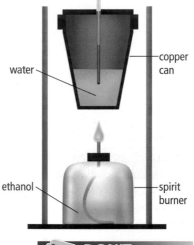

water — copper can

ethanol — spirit burner

DON'T FORGET

Enthalpy of combustion values must be calculated for one mole of the substance being burned.

LET'S THINK ABOUT THIS

25 cm³ of 1·0 mol l⁻¹ HCl(aq) was exactly neutralised by 50 cm³ of 0·5 mol l⁻¹ NaOH(aq). During the reaction the temperature rose by 4·5°C. Calculate the enthalpy of neutralisation of hydrochloric acid.

The correct answer is −56·43 $kJ\,mol^{-1}$.

To help you get this you may wish to look again at the definition for enthalpy of neutralisation.

You have to assume that all the solution is water and that the volume of water is the same as the total volume of solution. This enables you to work out the mass of water to be used in $E_h = cm\Delta T$.

You can calculate the number of moles of HCl and NaOH and therefore the number of moles of water formed. The calculation then has to be worked out for one mole of water being formed.

Since the reaction is exothermic, ΔH will be negative.

THE IDEA OF EXCESS

CALCULATING WHICH REACTANT IS IN EXCESS

What is meant by 'in excess'?

When a chemical reaction involving two reactants is carried out, usually one reactant gets used up completely and some of the other reactant is left over. The reactant left over is said to be 'in excess'. Sometimes it is necessary to calculate which reactant is in excess. The reactant which is not in excess will be used up and so it can be used to calculate how much product will be formed.

To work out which reactant is in excess it is necessary to calculate the number of moles of each reactant and use these values with the balanced formula equation.

Calculating the number of moles from mass of substance

This can be done using the formula triangle:

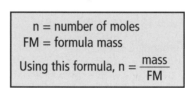

n = number of moles
FM = formula mass
Using this formula, $n = \dfrac{mass}{FM}$

DON'T FORGET

This is the formula to use if you are given the mass of a substance.

Example

Calculate the number of moles in 0·972 g of magnesium.

Answer

$$n = \frac{mass}{FM} = \frac{0.972}{24.3} = 0.040 \text{ moles}$$

Calculating the number of moles from volume and concentration

This can be done using another formula triangle:

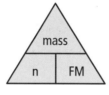

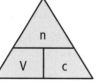

n = number of moles
V = volume of solution (in litres)
c = concentration of solution (in mol l^{-1})

DON'T FORGET

This is the formula to use if you are given the volume and concentration of a solution.

Example

Calculate the number of moles in 50 cm³ of 0·10 mol l^{-1} hydrochloric acid solution.

Answer

$V = 50 \text{ cm}^3 = 0.050 \text{ litres}$ $n = V \times c = 0.050 \times 0.1 = 0.0050 \text{ moles}$

Which reactant is in excess?

Which reactant would be in excess if 0·972 g of magnesium was added to 50 cm³ of 0·10 mol l^{-1} hydrochloric acid?

Using the results of the calculations above,
0·972 g of Mg is 0·040 moles and 50 cm³ of 0·1 mol l^{-1} HCl is 0·0050 moles.

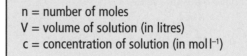

$$Mg(s) + 2HCl(aq) \longrightarrow MgCl_2(aq) + H_2(g)$$

The balanced formula equation for the reaction is given above and it shows that 1 mol of Mg will react with 2 mol of HCl.

So 0·040 mol of Mg would react with 0·080 mol of HCl.

Since there is only 0·0050 mol of HCl present and 0·080 mol is required then there is not enough HCl. The Mg is therefore in excess.

contd

CALCULATING WHICH REACTANT IS IN EXCESS contd

How much product will be formed?

The quantity of product formed can be calculated using the balanced formula equation and the number of moles of the reactant **which is not in excess**, that is the reactant which **is used up** completely.

Example

Calculate the mass of hydrogen produced when 0·972 g of magnesium is added to 50 cm³ of 0·10 mol l⁻¹ hydrochloric acid.

Answer

The balanced formula equation for the reaction is given below and it shows that 1 mol of Mg will react with 2 mol of HCl. The number of moles of each reactant has been calculated above and it has also been shown that magnesium is in excess. This means that the number of moles of HCl – the reactant which reacts completely – must be used in the calculation. This is shown below,

Balanced equation:	Mg(s)	+ 2HCl(aq) ⟶	MgCl$_2$(aq)	+ H$_2$(g)
From the equation:	1 mol	2 mol	1 mol	1 mol
From the number of moles of HCl calculated above:		0·0050 mol		0·0025 mol

(The equation shows that 1 mol of H$_2$ will be formed when 2 mol of HCl reacts completely so 0·0025 mol of H$_2$ will be formed when 0·005 mol of HCl reacts completely.)

Mass of hydrogen formed = n × FM = 0·0025 × 2 = **0·0050 g**.

Another calculation

Calculate the mass of carbon dioxide produced when 0·506 g of magnesium carbonate is added to 100 cm³ of 0·50 mol l⁻¹ nitric acid.

Firstly calculate the number of moles of each reactant.

The **mass** of magnesium carbonate is given, so $n = \dfrac{mass}{FM}$ The formula of magnesium carbonate is MgCO$_3$ and the formula mass works out at 84·3 So n for Mg $= \dfrac{0·506}{84·3} = 0·0060$ mol	The **volume** and **concentration** of nitric acid are given, so $n = V \times c$ The volume of nitric acid is 100 cm³ which is 0·10 litres. So n for HNO$_3$ = 0·10 × 0·50 = 0·050 mol

Then write the balanced chemical equation for the reaction:

$$MgCO_3(s) + 2HNO_3(aq) \longrightarrow Mg(NO_3)_2(aq) + H_2O(l) + CO_2(g)$$
$$\quad 1\text{ mol} \qquad 2\text{ mol}$$

The balanced equation tells us that 1 mole of MgCO$_3$ reacts with 2 moles of HNO$_3$.
Therefore 0·0060 mol of MgCO$_3$ would react with 0·012 mol (from 0·0060 × 2) of HNO$_3$.
Since there is 0·050 mol of HNO$_3$ present and only 0·012 mol will react, then HNO$_3$ is in excess.
This means that all the MgCO$_3$ will react.

Balanced equation:	MgCO$_3$(s)	+ 2HNO$_3$(aq) ⟶	Mg(NO$_3$)$_2$(aq)	+ H$_2$O(l)	+ CO$_2$(g)
From the equation:	1 mol	2 mol	1 mol	1 mol	1 mol

Therefore, 0·0060 mol of MgCO$_3$ will react to produce 0·0060 mol of CO$_2$.
The formula mass of CO$_2$ is 44, therefore, mass of CO$_2$ formed = n × FM = 0·0060 × 44 = 0·264 g.

⚙ LET'S THINK ABOUT THIS

Calculate
(i) The mass of copper(II) carbonate, CuCO$_3$, required to react completely with 100 cm³ of 2 mol l⁻¹ sulphuric acid.
(ii) The mass of carbon dioxide which would be produced when 1·34 g of copper(II) carbonate is added to 100 cm³ of 2 mol l⁻¹ sulphuric acid.
For answers, see p108.

PATTERNS IN THE PERIODIC TABLE

THE PERIODIC TABLE

Groups and Periods

The modern Periodic Table is based on the work of the Russian chemist, Dmitri Mendeleev (1839–1907), who arranged the elements known at that time in order of increasing relative atomic mass. He also took account of their chemical properties and put the elements with similar chemical properties together into the same group or vertical column. This was the start of the Periodic Table as we now know it. Nowadays, the elements are arranged in order of increasing atomic number.

The **vertical columns are known as groups** and the **horizontal rows are known as periods**.

All the elements in any one group have similar chemical properties. For example, all the elements in Group 1, the alkali metals, will react vigorously with water to produce hydrogen and an alkaline solution. The elements in Group 7, the halogens, are very reactive non-metals which react with metals to form salts.

Densities, melting points and boiling points

Values for **densities** of the elements are given on **page 3** of the **Data Booklet**. In general, the metals have greater densities than non-metals. However, looking at elements with atomic numbers 1–20, the element with the greatest density is carbon in the form of diamond with a density of $3.51\,g\,cm^{-3}$.

Melting points and boiling points of the elements are given on **page 4** of the **Data Booklet**.

Elements with melting points above room temperature (about 25°C) are solids at room temperature. Most elements are solid at room temperature.

Elements with boiling points below room temperature are gases at room temperature. All the elements in Group 0, the noble gases, have boiling points below room temperature.

Elements which have melting points below room temperature and boiling points above room temperature are liquids at room temperature. Mercury and bromine are the only two elements which are liquid at room temperature.

Covalent radius

The covalent radius of an element is half the distance between the nuclei of two of its bonded atoms. **The covalent radius** of each element given on **page 5** of the **Data Booklet** is taken to be a measure of the size of the atoms of that element.

As expected, the **covalent radius increases down a group**. This is because on moving down a group from one element to the next the number of occupied electron shells or energy levels increases.

The covalent radius across a period decreases from left to right. The atoms of all the elements in a period (horizontal row) have the same number of electron shells but **the nuclear charge is increasing**. The nuclear charge is related to the number of protons in the nucleus. Moving across a period from left to right there is an increase of one proton in the nucleus from one element to the next. This increase in nuclear charge exerts an increasing attraction on the outer electrons resulting in a decrease in covalent radius.

The covalent radius is an example of a **periodic property**, that is there is a definite pattern with increasing atomic number. This pattern is repeated across the different **periods** or rows in the Periodic Table.

Ionisation energy

The first ionisation energy of an element is the energy required to remove one mole of electrons from one mole of gaseous atoms of the element.

DON'T FORGET

Values for covalent radius, density, melting and boiling points of the elements are given in the Data Booklet.

contd

THE PERIODIC TABLE contd

The general equation representing the first ionisation energy is given at the top of page 10 in the Data Booklet and is $E(g) \longrightarrow E^+(g) + e^-$, in which E represents any element.

This means that the equation for the first ionisation energy for sodium is
$Na(g) \longrightarrow Na^+(g) + e^-$ and for chlorine, $Cl(g) \longrightarrow Cl^+(g) + e^-$.
The second and subsequent ionisation energies refer to the energies required to remove further moles of electrons.

For example, **the second ionisation energy refers to the equation, $E^+(g) \longrightarrow E^{2+}(g) + e^-$**, so the equation representing the second ionisation energy for potassium is $K^+(g) \longrightarrow K^{2+}(g) + e^-$.

The third ionisation energy refers to the equation, $E^{2+}(g) \longrightarrow E^{3+}(g) + e^-$, so the equation representing the third ionisation energy of aluminium is $Al^{2+}(g) \longrightarrow Al^{3+}(g) + e^-$.

Notice that in each equation representing an ionisation, **the reactant and product are both in the gaseous state** and that only one mole of electrons is removed.

$Al(g) \longrightarrow Al^{3+}(g) + 3e^-$ is the sum of the first, second and third ionisation energies of aluminium.

DON'T FORGET

The equation must show that the atom and the ion formed are in the gaseous state.

Trends in ionisation energies

Ionisation energy values are given on **page 10** of the **Data Booklet**. Values are given for the first, second, third and fourth ionisation energies.

Looking at the values of the first ionisation energies for the elements in the second period of the Periodic Table (Li to Ne), the **general trend** is that the **ionisation energy increases across a period from left to right**. Ionisation energy is also a periodic property as this same pattern is repeated from Na to Ar.

Since the ionisation energy is generally increasing **across a period**, more energy must be required to remove the outer electron. Reasons for this include:
- the positive nuclear charge is increasing so it is more difficult for the negative electron to be 'pulled' away,
- the atoms are getting smaller and so the negative electron is closer to the positive nucleus and attracted more strongly.

DON'T FORGET

At each stage one mole of electrons is removed. This is why the units are kJ mol^{-1}.

Looking at the values for lithium, sodium and potassium or the first three elements in any group you can see that the **ionisation energy decreases down a group in the Periodic Table**. This means that less energy is required to remove the outer electron. Reasons for this include:
- the outer electron to be removed is in a **shell further away** from the nucleus so the atoms are getting bigger and so the negative electron is further away from the positive nucleus and attracted less strongly,
- there will a **greater screening effect** due to more shells of inner electrons between the outer shell and the nucleus. These inner electrons are said to 'screen' or 'shield' the electrons in the outer shell from the positive nucleus.

Electronegativity

Atoms of different elements have different attractions for bonding electrons. Atoms of non-metal elements usually, but not always, have a greater attraction for electrons within a bond than atoms of metal elements.

Electronegativity is a measure of the attraction an atom has for the electrons in a bond.

Electronegativity values of most elements are given on page 10 of the Data Booklet. Notice that the values have no units. **Electronegativity values increase along a period and decrease down a group**. Fluorine has the greatest electronegativity and so is the element which has the greatest attraction for the electrons in a bond. Alkali metals have the least attraction for the electrons in a bond and so have the lowest electronegativity values.

LET'S THINK ABOUT THIS

Explain (i) why there are no electronegativity values given for the Noble Gases (Group 0)
(ii) why there is no fourth ionisation energy value given for lithium
(iii) the large 'jump' between the third and fourth ionisation energy values for aluminium.
For answers, see p108.

BONDING, STRUCTURES AND PROPERTIES 1

TYPES OF BONDING

Metallic bonding

Metallic elements are at the left side of the Periodic Table and generally have low ionisation energies compared to non-metals. This means that they lose their outer electrons more readily to form positive ions. Atoms in a metal contribute their outer electrons to a 'pool' of freely moving or **delocalised electrons**. The structure of metals in the solid state is regarded as a regular array or lattice of positive ions held together by a 'sea' of delocalised electrons.

> **DON'T FORGET**
>
> Metallic bonding is the electrostatic force of attraction between positively charged metal ions and negatively charged delocalised outer electrons.

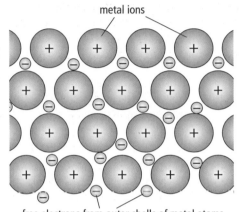

metal ions

free electrons from outer shells of metal atoms

The **metallic bond is the electrostatic attraction between the positive ions and the negative delocalised electrons** and this holds the metallic structure together as a single unit. The delocalised outer electrons are free to move throughout the metallic structure or lattice. The presence of these delocalised electrons explains why metals are very good conductors of heat and electricity. Other properties of metals explained by the delocalised electrons are:
- ductility – can be drawn into wires
- malleability – can be beaten into sheets.

Metallic bonding is fairly strong and so metals generally have high melting and boiling points. An exception is mercury. The strength of the metallic bond depends to some extent on the number of outer electrons. The alkali metals (Group 1) have only one outer electron and have lower melting and boiling points than most other metals.

Covalent bonding

Covalent bonds are formed by the merging or overlapping of half-filled outer electron clouds between the positive nuclei of two atoms. The positive nuclei of both atoms attract the electrons in the overlap region and this is what holds the atoms together.

> **DON'T FORGET**
>
> Atoms in a covalent bond are held together by electrostatic forces of attraction between positively charged nuclei and negatively charged shared electrons.

Covalent bonds are usually, but not always, formed between atoms of non-metal elements. A molecule is a group of atoms held together by covalent bonds. Covalent bonds are strong bonds and a great deal of energy must be put in to break the covalent bonds inside molecules.

If two atoms that are linked together by covalent bonding are from the same element or have the **same electronegativity value**, then they have an equal 'pull' on the shared electrons and the bonding is pure covalent or **non-polar covalent**. Elements which exist as diatomic molecules such as hydrogen or chlorine have non-polar covalent bonding. The compound PH_3 has non-polar covalent bonding since both phosphorus and hydrogen have the same electronegativity value.

Polar covalent bonding

When the two atoms linked by covalent bonding are from elements with different electronegativity values then there will be an unequal 'pull' on the shared electrons. The atom of the element with the greater electronegativity value will have the greater share of the electrons and it will take on a slight negative charge. The other atom will have a slight positive charge. Effectively this means that there is an **unequal sharing** of the bonding electrons and this is known as **polar covalent** bonding.

contd

TYPES OF BONDING contd

In hydrogen chloride, chlorine has a greater electronegativity than hydrogen. Therefore, chlorine has a greater 'pull' on the shared electrons than hydrogen. This is shown as $\overset{\delta+}{H}$—$\overset{\delta-}{Cl}$ in which δ (delta) means 'slightly' and this shows that the bond is polar, with the hydrogen atom having a permanent slight positive charge and the chlorine atom having a permanent slight negative charge.

The degree of polarity of a covalent bond depends on the difference in electronegativity between the bonded atoms. The greater the difference in electronegativity then the more polar the covalent bond. So an H—O bond is more polar than an H—Cl bond.

Ionic bonding

When a metal element from the left side of the Periodic Table joins with a non-metal element from the right of the Periodic Table the type of bonding is usually ionic. As a very rough guide, if the difference in electronegativity is greater than 1·5 the bonding is usually, but not always, ionic.

When an ionic bond forms, the loosely held outer electron(s) of the metal atom are attracted into the outer shell of the non-metal atom, forming positive and negative ions, respectively. Ionic bonding is the electrostatic force of attraction between these positive and negative ions. The ions are held together in an ionic lattice. An exploded diagram of the ionic lattice of sodium chloride is shown below.

DON'T FORGET

The greater the difference in electronegativity between the two elements then the more likely the bonding is to be ionic. If the electronegativity of two non-metal elements is the same, then the bond between them will be non-polar covalent.

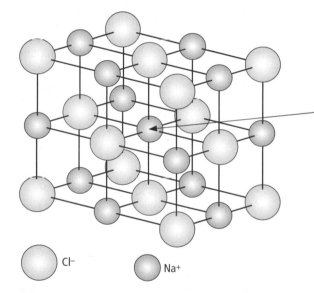

The ions with lines showing between them in this exploded diagram will actually be 'touching' and it can be seen in this diagram that the positive sodium ion indicated is 'touching' six negative chloride ions but no ions of the same charge are 'touching'.

Cl⁻   Na⁺

The type of bonding in a compound is related to the relative positions of its elements in the Periodic Table. The closer the elements are to each other then the more likely the bonding is to be covalent. The further away they are from each other then the more likely the bonding is to be ionic.

LET'S THINK ABOUT THIS

The classification of bonds into covalent or ionic is an over simplification. Pure covalent and pure ionic are at opposite ends of a continuum with the bonding in most compounds being somewhere in between. Electronegativity differences are useful predictors of the type of bonding, but only by examining the properties of the compounds can the type of bonding be identified conclusively. The main properties of ionic compounds are that they have high melting and boiling points and conduct electricity only when molten and in solution.

Use your Data Booklet to find the electronegativity difference between the elements in water, H_2O, and in sodium hydride, NaH. One compound is a liquid at room temperature, the other is a solid which conducts electricity when molten. Which of these compounds has ionic bonding and write its ionic formula?

For answer, see p108.

BONDING, STRUCTURES AND PROPERTIES 2

INTERMOLECULAR FORCES OF ATTRACTION

Van der Waals' forces

The word 'intermolecular' means **between** molecules and does not mean inside molecules. Intermolecular forces of attraction are very weak forces between different molecules and sometimes between individual discrete atoms. Metallic bonds, covalent bonds and ionic bonds are much stronger than intermolecular forces.

The weakest of all the intermolecular forces are van der Waals' forces. They are most evident when molecular elements such as hydrogen, oxygen, nitrogen and so on, are liquefied on cooling. Van der Waals' forces are the only forces of attraction between these molecules. Without van der Waals' forces these molecular elements would exist as gases at even the lowest temperatures. This is also true for the noble gases. These elements exist as monatomic gases but due to van der Waals' forces they can also be liquefied at low temperatures.

Van der Waals' forces can operate **between** all atoms and molecules, and are caused by the movement of electrons inside atoms and molecules. At any one instant the electron distribution in an atom or molecule might be such that there is a very slight negative charge at one side of the atom or molecule. There will be a corresponding very slight positive charge at the other side. This unequal distribution of charge means that a **temporary dipole** has been created. The positive end of a dipole on one atom or molecule is attracted to the negative end of a dipole on a nearby atom or molecule.

If one end of an atom or molecule is slightly negative it will repel negative electrons on adjacent atoms or molecules causing dipoles there. Temporary dipoles formed this way are said to be induced dipoles.

Van der Waals' forces are a result of electrostatic attraction between temporary dipoles and/or induced dipoles caused by the movement of electrons in atoms and molecules.

The strength of van der Waals' forces is proportional to the size of the atoms or molecules. The **bigger the atoms or molecules then the stronger the van der Waals' forces** between these atoms or molecules. These van der Waals' forces have to be overcome before elements which exist as molecules or as single atoms will melt or boil. This explains why there is a definite increase in the boiling points of the elements moving down Group 7 and Group 0 as shown in the table below.

DON'T FORGET

Van der Waals' forces are caused by temporary dipoles, are the weakest intermolecular force and depend on the size of the atoms or molecules.

Element	Boiling point (°C)
Fluorine, F_2	−188
Chlorine, Cl_2	−35
Bromine, Br_2	59
Iodine, I_2	184

Element	Boiling point (°C)
Helium, He	−269
Neon, Ne	−246
Argon, Ar	−186
Krypton, Kr	−152

Polar molecules and permanent dipole–permanent dipole attractions

If a molecule has polar covalent bonds and has a **permanent dipole** it is said to be a **polar molecule**. Compounds with polar molecules include hydrogen chloride, HCl, water, H_2O, and ammonia, NH_3. Chlorine is more electronegative than hydrogen and so in HCl, the hydrogen atom is slightly positive ($\delta+$) and the chlorine atom is slightly negative ($\delta-$).

However, not all molecules with polar covalent bonds are polar molecules. As well as the polarity of the bonds the **spatial arrangement or shape** of the molecule must also be taken into account.

contd

INTERMOLECULAR FORCES OF ATTRACTION contd

If the molecule is symmetrical the **polarity of the bonds may cancel out** and the molecule itself may be non-polar. For example, carbon dioxide and carbon tetrachloride (tetrachloromethane) have polar covalent bonds but because of their molecular shape and symmetry the polarity cancels out and overall they are non-polar molecules.

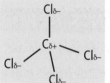

$$\overset{\delta-}{O} = \overset{\delta+}{C} = \overset{\delta-}{O}$$

The polarities of both C=O bonds cancel out and overall CO_2 is non-polar.

The polarities of the four C–Cl bonds cancel out and overall CCl_4 is a non-polar molecule.

The intermolecular forces between **non-polar molecules** are the weak **van der Waals' forces**. However, since polar molecules have a permanent dipole, the main intermolecular force between polar molecules are **permanent dipole–permanent dipole interactions**. These are electrostatic forces of attraction between a slightly positive atom of one molecule and a slightly negative atom in an adjacent atom.

For example, the compound methanal, CH_2O, has a polar C=O bond. Its molecular structure is shown below. The slightly negative oxygen atoms in one methanal molecule will be attracted to the slightly positive carbon atom in another methanal molecule. This is a permanent dipole–permanent dipole attraction since the oxygen atom is always slightly negative in methanal and the carbon atom is always slightly positive. The permanent dipole–permanent dipole attraction is shown as a dotted line and the covalent bonds are shown as solid lines.

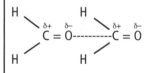

Permanent dipole–permanent dipole interactions are **stronger than van der Waals' forces** for molecules of equivalent size.

Hydrogen bonding

Bonds consisting of a hydrogen atom bonded to an atom of a strongly electronegative element such as fluorine, oxygen or nitrogen are highly polar. Therefore H–F, H–O and H–N bonds are highly polar. Compounds such as hydrogen fluoride, HF, water, H_2O, and ammonia, NH_3 contain these highly polar bonds. In molecules of these compounds the hydrogen atoms have a permanent δ+ charge. The slightly positive hydrogen atom in one molecule will be strongly attracted to the slightly negative fluorine, oxygen or nitrogen in adjacent molecules. These attractions are known as **hydrogen bonds**. Hydrogen bonds are electrostatic forces of attraction between molecules which contain these highly polar bonds: H–F, H–O or H–N. Since there is a permanent dipole present – the hydrogen atom is always slightly positive and the F, O or N atom is always slightly negative in these bonds – these attractions are a special case of permanent dipole–permanent dipole interactions. Hydrogen bonds are **stronger than other forms of permanent dipole–permanent dipole interactions** but much weaker than covalent bonds.

A water molecule, H_2O contains two highly polar O–H bonds. Hydrogen bonds will form between a hydrogen atom in one water molecule and an oxygen atom in a neighbouring water molecule as shown below.

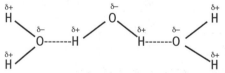

The hydrogen bonds are shown as dotted lines. The polar covalent bonds which are stronger are shown as solid lines.

DON'T FORGET

Hydrogen bonds are the strongest of the intermolecular forces and van der Waals' forces are the weakest.

LET'S THINK ABOUT THIS

1   Which of these substances will have hydrogen bonds between their molecules: C_2H_5OH, HCl, CH_3NH_2, HF, H_2, H_3COCH_3? (It may help to draw their molecular structure first. Remember each C must have four bonds, each N three bonds and each O two bonds).
2   Which of the following substances will have polar molecules: HCl, CH_3Cl, $CHCl_3$, CF_4, H_2S, CS_2, NH_3, O_2? (It may help to draw out the molecular structures first).

For answers, see p108.

BONDING, STRUCTURES AND PROPERTIES 3

ELEMENTS 1–20

Metallic

A metallic structure consists of a giant lattice of positively charged ions and delocalised outer electrons (see diagram, page 22). Of the first 20 elements in the Periodic Table, **Li, Be, Na, Mg, Al, K and Ca** have a metallic structure and metallic bonding.

Covalent molecular

A covalent molecular structure consists of discrete molecules held together by weak intermolecular forces. In the case of elements with covalent molecular structures, these intermolecular forces are the very weak van der Waals' forces. Elements 1–20 with **covalent molecular bonding** are:

- **hydrogen**, H–H; **nitrogen**, N≡N; **oxygen**, O=O; **fluorine**, F–F; and **chlorine**, Cl–Cl
- **phosphorus**, with P_4 molecules
- **sulphur**, with S_8 molecules
- **carbon**, in its fullerene form with molecules such as C_{60}, C_{70}, C_{240}, C_{360} and so on.

The molecular structures of P_4, S_8 and C_{60} are shown on the left.

Elements which have covalent molecular structures have **low melting and boiling points** since there are only **very weak van der Waals' forces between the molecules**. Some of these must be overcome at the melting point when the solid becomes a liquid and all of them have to be overcome at the boiling point when the liquid changes to a gas. Van der Waals' forces between bigger molecules are slightly stronger and this is why phosphorus, sulphur and the fullerenes are solid at room temperature. The molecules, themselves, are still the same in the solid, liquid and gas states so the strong **covalent bonds** inside the molecules are **not broken** at the melting or boiling points of the elements.

Covalent network

A covalent network structure consists of a giant lattice of covalently bonded atoms. There are three elements in the first 20 which have covalent network structures. These are **boron, silicon and carbon as diamond or graphite**. These elements have **very high melting and boiling points** since **covalent bonds** which are very strong have to be broken at their melting and boiling points.

Boron is a non-metal with a covalent network structure based on interlocking B_{12} units, making a structure containing many billions of boron atoms joined together.

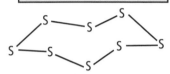

Each P atom has one bond to each of the other three P atoms.

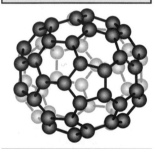

Each S atom has one bond to two other S atoms.

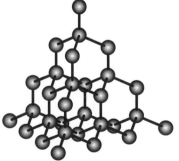

Each C atom has one bond to two other C atoms.

Research is being conducted to try to find uses for fullerenes.

Silicon and diamond (a pure form of carbon) have similar structures with each atom covalently bonded to another four atoms in a gigantic tetrahedral arrangement of atoms.

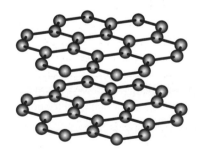

Part of the structure of graphite, another pure form of carbon. The carbon atoms form layers of hexagonal rings. Each carbon atom is joined covalently to another three carbon atoms. The layers of rings are held together by weak van der Waals' forces. Each carbon atom has a spare electron which becomes delocalised. This explains why graphite is a good conductor of electricity.

Monatomic structures

The noble gases, Group 0, exist as single atoms (monatomic). The noble gases all have a very stable outer electron shell. Among the first 20 elements, those which exist as monatomic gases are helium, He, neon, Ne, and argon, Ar.

There is no bonding involved except for the very weak van der Waals' forces between the atoms when they are close together in the liquid and solid states. These forces are easily overcome and so the noble gases have very low melting and boiling points and are all gases at room temperature.

contd

COMPOUNDS

Covalent molecular

Many covalent compounds are made up of small discrete individual molecules. Inside a molecule the atoms are covalently bonded to each other. Typical covalent molecular compounds include CO_2, H_2O and CCl_4. Hydrocarbons and carbohydrates have covalent molecular structures.

The melting and boiling points of polar compounds are higher than the melting and boiling points of non-polar substances with similar molecular sizes. This is because the polar compounds have the stronger intermolecular permanent dipole-permanent dipole interactions, compared to non-polar compounds and elements which have only the very weak van der Waals' forces between their molecules.

Covalent network

Some compounds have a covalent network structure. These include silicon carbide and silicon dioxide. Their formulae are written SiC and SiO_2, respectively. These are empirical formulae rather than molecular formulae since they contain many millions of atoms covalently bonded together in a large network structure. They do not contain individual discrete molecules.

These compounds have **very high melting and boiling points**, and like diamond, which is a covalent network element, these compounds are also **very hard**. Silicon carbide, for example, is used as an abrasive.

Ionic compounds

An ionic structure consists of a giant lattice of oppositely charged ions. (See diagram of the ionic lattice of sodium chloride, page 23.)

The electrostatic attractions between ions within the ionic lattice are very strong. These attractions must be overcome so that the lattice is broken when the compound melts. This requires a lot of energy and this is why ionic compounds are always solid at room temperature. Ionic compounds have high melting points and boiling points because of the strong attractions between the oppositely charged ions.

Solubility

In general, the more polar the substance is then the more likely it is to dissolve in a polar solvent such as water and the less likely it is to dissolve in a non-polar solvent such as hexane or carbon tetrachloride, CCl_4.

Non-polar molecular substances are usually soluble in non-polar solvents such as hexane or CCl_4 and tend to be insoluble in water and other polar solvents.

Properties associated with hydrogen bonding

Covalent molecular compounds which have hydrogen bonds between their molecules, that is those containing polar H–O, H–N and H–F bonds, will have higher melting and boiling points than other covalent molecular substances of similar molecular size because it requires more energy to break hydrogen bonds than other intermolecular forces. For example, hydrogen fluoride, water and ammonia all have much higher melting and boiling points than would be expected from their small molecular sizes.

Hydrogen bonding between molecules also means that the compound is more likely to have a higher viscosity or be viscous (a thick liquid).

Compounds which form hydrogen bonds to water molecules are more likely to be soluble in water.

When water freezes into ice, the water molecules move slightly apart to allow for the maximum number of hydrogen bonds to form. This explains why ice floats on water since it is less dense than water. In fact, water is at its maximum density at 4°C.

LET'S THINK ABOUT THIS

Which bonds or intermolecular forces are being broken when the following substances melt: silicon dioxide, water, sulphur, zinc, sodium oxide, ammonia, phosphorus, aluminium oxide, phosphorus hydride? For answer, see p108.

THE MOLE 1

THE AVOGADRO CONSTANT

The mole as a number

A **mole** of any substance is the **formula mass in grams** of that substance. For example, the formula mass of water is 18·0 and so one mole of water has a mass of 18·0 g.

However, the mole can also be expressed as a number. One mole of any substance contains $6·02 \times 10^{23}$ **formula units**. This number is known as the **Avogadro Constant**. It has the symbol, **L**, and its value is given on page 19 of the Data Booklet.

One mole of any substance contains the same number of formula units.

1 mole of sodium	1 mole of sulphur	1 mole of water	1 mole of calcium chloride

23·0 g of Na	**32·1 g of S**	**18·0 g of H_2O**	**111·0 g of $Ca^{2+}(Cl^-)_2$**

Each of the substances above contains 6×10^{23} formula units.

Equimolar (same number of moles) amounts of substances contain equal numbers of formula units. Therefore 0·1 mol of any substance will contain $6·02 \times 10^{23} \times 0·1 = 6·02 \times 10^{22}$ formula units. 2·0 mol of any substance contains $2 \times 6·02 \times 10^{23} = 1·204 \times 10^{24}$ formula units.

The number of formula units = nL, where n is the number of moles, and L is the Avogadro Constant.

Formula units

Formula units can be atoms, molecules or groups of ions. How the formula is written and the type of structure must be considered. Substances which have **atoms** as their formula units are those for which the **chemical formula** is just the chemical **symbol**. The table below gives examples of some of these substances.

Substance	Type of structure	Formula	Formula units
Sodium	Metallic	Na	Na atoms
Aluminium	Metallic	Al	Al atoms
Copper	Metallic	Cu	Cu atoms
Helium	Monatomic	He	He atoms
Argon	Monatomic	Ar	Ar atoms
Phosphorus	Covalent molecular	P	P atoms
Sulphur	Covalent molecular	S	S atoms
Diamond	Covalent network	C	C atoms

Substances which have **molecules** as their formula units have a covalent molecular structure and their **chemical formulae** are identical to their **molecular formulae**. The table below gives examples.

Substance	Type of structure	Formula	Formula units
Hydrogen	Covalent molecular	H_2	H_2 molecules
Water	Covalent molecular	H_2O	H_2O molecules
Carbon dioxide	Covalent molecular	CO_2	CO_2 molecules
Glucose	Covalent molecular	$C_6H_{12}O_6$	$C_6H_{12}O_6$ molecules

Ionic compounds have groups of ions as their formula units. Ionic compounds do not contain atoms or molecules. The positive and negative ions are held together in an ionic lattice. The formula unit is the smallest group of positive ions and negative ions required to balance the ion charges. The table on the opposite page shows examples.

DON'T FORGET

- If the structure is **metallic**, the formula units are **atoms**.
- The formula units for the **noble gases** are also **atoms**.
- Even though sulphur and phosphorus exist as S_8 and P_4 molecules, respectively, their formula units are written as S and P atoms.

DON'T FORGET

If the chemical formula is written as the molecular formula, then the formula units are molecules.

contd

THE AVOGADRO CONSTANT contd

Substance	Type of structure	Formula	Formula units
Sodium chloride	Ionic	NaCl	Na^+Cl^-
Calcium oxide	Ionic	CaO	$Ca^{2+}O^{2-}$
Calcium chloride	Ionic	$CaCl_2$	$Ca^{2+}(Cl^-)_2$

Using the Avogadro Constant

1 mole of any substance contains 6.02×10^{23} formula units.
Using some of the examples in the tables above, then
$23.0\,g$ (1 mol) of sodium contains 6.02×10^{23} Na atoms
$63.5\,g$ (1 mol) of copper contains 6.02×10^{23} Cu atoms
$44.0\,g$ (1 mol) of carbon dioxide contains 6.02×10^{23} CO_2 molecules
$58.5\,g$ (1 mol) of sodium chloride contains 6.02×10^{23} (Na^+Cl^-) formula units.

Worked examples:

1 Calculate the number of **atoms** in $5.4\,g$ of aluminium.

Answer:

First calculate the number of moles, n, of aluminium.

$n = \dfrac{mass}{FM} = \dfrac{5.4}{27.0} = 0.2$ mol

The number of formula units (FU) = n × L = $0.2 \times 6.02 \times 10^{23} = 1.204 \times 10^{23}$ FU.
For aluminium, the formula units are atoms, so **the number of atoms = 1.204×10^{23}**.

2 Calculate the number of **molecules** in $0.176\,g$ of carbon dioxide.

Answer:

The number of moles, n, of $CO_2 = \dfrac{mass}{FM} = \dfrac{0.176}{44.0} = 0.004$ mol.

The number of formula units = n × L = $0.004 \times 6.02 \times 10^{23} = 2.408 \times 10^{21}$ FU.
For carbon dioxide, the formula units are molecules, so the **number of molecules = 2.408×10^{21}**.

3 Calculate the number of **atoms** in $100.0\,g$ of water.

Answer:

$n = \dfrac{mass}{FM} = \dfrac{100.0}{18.0} = 5.56$ mol

The number of formula units = n × L = $5.56 \times 6.02 \times 10^{23} = 3.35 \times 10^{24}$ FU.
For water, the formula units are molecules, so the number of molecules = 3.35×10^{24} molecules.
However, the question asks for the **number of atoms**. Each water molecule contains three atoms (two hydrogen atoms + one oxygen atom).
So the number of atoms = $3.35 \times 10^{24} \times 3 = 1.01 \times 10^{25}$.

4 Calculate the number of ions in $2.22\,g$ of calcium chloride.

Answer:

Calcium chloride has chemical formula $CaCl_2$. Its ionic formula is $Ca^{2+}(Cl^-)_2$ and it has formula mass 111.

$n = \dfrac{mass}{FM} = \dfrac{2.22}{111} = 0.02$ mol

The number of formula units = n × L = $0.02 \times 6.02 \times 10^{23} = 1.204 \times 10^{22}$ FU.
The formula unit is $Ca^{2+}(Cl^-)_2$. Each formula unit has three ions (1 × Ca^{2+} and 2 × Cl^- ions).
So the number of ions = $1.204 \times 10^{22} \times 3 = 3.61 \times 10^{22}$ ions.

DON'T FORGET

There are no molecules or atoms in ionic compounds, only positive and negative ions.

DON'T FORGET

Calculate the number of moles, n, first, then use n × L to get the number of formula units.

LET'S THINK ABOUT THIS

1 Calculate the number of **atoms** in $1\,g$ of (a) helium (b) methane, CH_4 (c) ethanol, C_2H_5OH.
2 Calculate the number of **ions** in $1\,g$ of
(a) sodium chloride, NaCl (b) calcium carbonate, $CaCO_3$ (c) ammonium sulphate, $(NH_4)_2SO_4$.
3 Calculate the number of **hydrogen atoms** in $5\,g$ of
(a) hydrogen, H_2 (b) ammonia, NH_3 (c) hexane, C_6H_{14} (d) glucose, $C_6H_{12}O_6$.
For answers, see p108.

THE MOLE 2

MOLAR VOLUME AND REACTING VOLUMES

Molar volume?

The mass of one mole of different substances is likely to be different. Although the mass may be different, **the volume of one mole of different gases is the same under the same conditions of temperature and pressure**.

For example, this means that one mole of hydrogen weighing 2·0 g will occupy the same volume as one mole of sulphur dioxide weighing 64·1 g at the same temperature and pressure.

It is also true to state that 0·20 g of hydrogen (0·1 mol) would have the same volume as 6·41 g of sulphur dioxide (0·1 mol) at the same temperature and pressure.

The volume of one mole of gas is known as the **molar volume**.

At room temperature, which is about 25°C, and 1 atmosphere pressure, the molar volume of any gas is about 24 litres.

This also means that **equal volumes of all gases** at the same temperature and pressure will have the **same number of formula units**. For the Noble Gases the formula units are atoms. For all other gases, the formula units are molecules.

Using the molar volume in calculations

The fact that all gases have the same molar volume allows us to calculate the volume of a gas from the number of moles and *vice-versa*.

In the examples below, the molar volume of gases at room temperature and pressure will be taken to be 24·0 litres mol⁻¹. (Note the units, **litres per mole**.)

Example 1

Calculate the volume of 4 moles of carbon dioxide at room temperature and pressure.

Answer:
1 mol of carbon dioxide has a volume = 24·0 litres.
Therefore, 4 moles will have a volume = 24·0 × 4 = **96·0 litres**.

Example 2

Calculate the volume of 1·0 g of helium at room temperature and pressure.

Answer:
Firstly calculate the number of moles of helium, $n = \dfrac{mass}{FM} = \dfrac{1·0}{4·0} = 0·25$ mol.
Volume of 1 mol = 24·0 litres.
Volume of 0·25 mol = 0·25 × 24·0 = **6·0 litres**.

Example 3

Calculate
(a) the number of moles in 0·60 litres of oxygen
(b) the mass of 0·60 litres of oxygen.

Answer:
(a) The molar volume = 24·0 litres mol⁻¹, therefore the number of moles, $n = \dfrac{0·60}{24·0} = 0·025$ mol.
(b) mass = n × FM = 0·025 × 32·0 = **0·8 g**

DON'T FORGET

The formula mass of the monatomic gases is the same as their relative atomic mass, but remember that for the diatomic elements their FM = RAM × 2.

contd

MOLAR VOLUME AND REACTING VOLUMES contd

Reacting volumes

As stated earlier, one mole of any **gas** contains the same number of formula units and has the same volume under the same conditions of temperature and pressure.

This means that the volumes of reactant and product **gases** involved in a chemical reaction can be calculated.

A balanced chemical equation gives the relative number of moles of each reactant and product substance. Therefore it also gives the **relative volumes** of any reactant and product substances that are **gases**.

Consider the reaction of sulphur dioxide and oxygen producing sulphur trioxide. The reactants and product are all gases and the balanced chemical equation is:

	$SO_2(g)$	+	$\frac{1}{2} O_2(g)$	⟶	$SO_3(g)$
This gives the mole ratio	1 mol		0·5 mol		1 mol
The ratio of reacting volumes is	1 volume		0·5 volume		1 volume
Therefore	1 litre		0·5 litres		1 litre
or keeping the same ratio	10 cm³		5 cm³		10 cm³

Note that any reactant and/or product substances that are solid or liquid have negligible volume compared to any gases present and in these calculations these substances are assumed to have zero volume.

For example, consider the reaction of hydrogen and oxygen producing water. The balanced equation is:

	$H_2(g)$	+	$\frac{1}{2} O_2(g)$	⟶	$H_2O(l)$
This gives the mole ratio	1 mol		0·5 mol		1 mol
The ratio of reacting volumes is	1 volume		0·5 volume		0 volume
Therefore	1 litre		0·5 litres		0 litre
or	10 cm³		5 cm³		0 cm³

Look at the worked examples below.

> **DON'T FORGET**
>
> Solids and liquids in the equation are assumed to have zero volume compared to any gases in the equation.

Example 1

(a) Calculate the volume of oxygen required for the complete combustion of 100 cm³ of methane.
(b) Calculate the volume of carbon dioxide produced.
To answer parts (a) and (b), start by writing the balanced chemical equation,

	$CH_4(g)$	+	$2O_2(g)$	⟶	$CO_2(g)$	+	$2H_2O(l)$
This gives the mole ratio	1 mol		2 mol		1 mol		2 mol
The ratio of reacting volumes is	1 volume		2 volumes		1 volume		0 volume
Therefore	100 cm³		200 cm³		100 cm³		

So the answer to (a) is 200 cm³ of oxygen and the answer to (b) is 100 cm³ of carbon dioxide.

🛠 LET'S THINK ABOUT THIS

To get a hydrocarbon to burn completely to produce carbon dioxide rather than carbon monoxide, it is necessary for oxygen to be in excess. Try this example.
200 cm³ of ethene gas (C_2H_4) was mixed with 900 cm³ of oxygen and the mixture ignited.

Calculate
(a) the volume of excess oxygen,
(b) the total volume of gas at the end of the experiment.

(Assume all measurements of gas volume are made under the same conditions of temperature and pressure.)

For answers, see p108.

PPA 1–3

PPA 1 – THE EFFECT OF CONCENTRATION CHANGES ON REACTION RATE

Introduction

In acidic conditions, hydrogen peroxide reacts with the iodide ions from potassium iodide to form water and iodine. The equation without spectator ions is:

Equation (1) $H_2O_2(aq) + 2H^+(aq) + 2I^-(aq) \longrightarrow 2H_2O(l) + I_2(aq)$

Starch and sodium thiosulphate solutions were also present.

The thiosulphate ions react with the iodine formed in reaction (1) and prevents it reacting with the starch.

Equation (2) $I_2(aq) + 2S_2O_3^{2-}(aq) \longrightarrow 2I^-(aq) + S_4O_6^{2-}(aq)$

When all the thiosulphate ions have been used up in reaction (2), the iodine formed in reaction (1) can then react with the starch to produce the familiar blue-black colour.

The number of moles of thiosulphate ions was the same each time and so the sudden appearance of the blue-black colour always represented the same extent of the reaction, that is when the same quantity of iodine had been formed each time.

Aim

To find out how changing the concentration of potassium iodide affects the reaction rate when potassium iodide reacts with hydrogen peroxide.

Procedure

Solutions of sulphuric acid, sodium thiosulphate, starch and potassium iodide were measured into a beaker.

The stopwatch was started when hydrogen peroxide was added to the same beaker and stopped as soon as the blue-black colour appeared.

In the experiment the concentration of potassium iodide was changed by diluting it with water but keeping the same total volume each time, as shown in the results table.

The procedure was repeated four times with different concentrations of potassium iodide.

The time, t, for the blue-black colour was measured in seconds and the reaction rate was taken to be $\frac{1}{t}$.

> **DON'T FORGET**
>
> The reaction rate is taken to be $\frac{1}{time}$ and the units are s^{-1}.

Results

Typical results are shown in the table below:

Volume of potassium iodide solution (cm³)	Volume of water (cm³)	Reaction time (s)	Rate (s⁻¹)
25	0	6·3	0·16
20	5	7·7	0·13
15	10	10·1	0·099
10	15	15·3	0·065
5	20	28·6	0·035

contd

PPA 1 – THE EFFECT OF CONCENTRATION CHANGES ON REACTION RATE contd

The results were plotted onto a graph and the graph produced was a straight line as shown on the right.

Conclusion

As the concentration of potassium iodide increases the reaction rate increases.

Evaluation

- Factors which had to be kept constant were temperature and the concentrations and volumes of the other reactants (sulphuric acid, sodium thiosulphate and starch).

- The total volume was kept the same in each experiment so that the concentrations of the other reactants were constant. The only factor changed in the experiment was the concentration of the potassium iodide solution.

- The procedure used in this experiment was effective because it was easy to measure the time taken for the blue colour to appear since the colour change from colourless to blue-black was very sharp.

- The reaction mixtures were placed on a white tile in order to make it easier to detect the colour change.

- Both sulphuric acid and hydrogen peroxide irritate the eyes and goggles had to be worn.

> **DON'T FORGET**
>
> The total volume of the reaction mixture was kept the same in each experiment. This keeps the concentrations of the hydrogen peroxide and sulphuric acid constant, and allows us to take **the volume of potassium iodide solution used to be a measure of the concentration of iodide ions.**

PPA 2 – THE EFFECT OF TEMPERATURE CHANGES ON REACTION RATE

Introduction

Oxalic acid reacts with an acidified solution of potassium permanganate. The equation without showing $K^+(aq)$ and $SO_4^{2-}(aq)$ ions which are spectator ions is:

$$5(COOH)_2(aq) + 6H^+(aq) + 2MnO_4^-(aq) \longrightarrow 2Mn^{2+}(aq) + 10CO_2(g) + 8H_2O(l)$$

Initially the reaction mixture is purple due to the presence of the permanganate, $MnO_4^-(aq)$ ions but it turns colourless as soon as they are used up. The amount of permanganate ions present is the same for all the experiments. This means that the point at which the purple colour disappears represents the same extent of reaction.

The time taken for the purple colour to disappear, t, is noted and the reaction rate is taken to be $\frac{1}{t}$. If the time is measured in seconds then the units will be s^{-1}.

> **DON'T FORGET**
>
> Reaction rate is taken to be $\frac{1}{\text{time}}$ and the units of reaction rate are s^{-1} if the time is measured in seconds.

Aim

To find the effect of changing temperature on the rate of the reaction between oxalic acid and acidified potassium permanganate.

Procedure

$5 \, cm^3$ of $1 \, mol \, l^{-1}$ sulphuric acid, $2 \, cm^3$ of potassium permanganate solution and $40 \, cm^3$ of water were measured out into a $100 \, cm^3$ **dry** glass beaker.

The mixture was heated to about 40°C and then the beaker was placed on a white tile and $1 \, cm^3$ of oxalic acid was added. The timer was started at the same time as the oxalic acid was added and the time taken for the mixture to change from purple to colourless was noted. The experiment was repeated at approximately 50°C, 60°C and 70°C.

contd

PPA 2 – THE EFFECT OF TEMPERATURE CHANGES ON REACTION RATE contd

> **DON'T FORGET**
>
> The graph is **not** a straight line graph. It curves upwards with increasing temperature. This shows that only a small increase in temperature produces a large increase in reaction rate.

Results

Typical results are shown in the table opposite:

These results were plotted onto a graph and this time the line curved upwards as shown below:

Temperature (°C)	Time (s)	Reaction rate (s⁻¹)
39	90·1	0·011
47	32·6	0·031
60	11·1	0·090
69	4·1	0·24

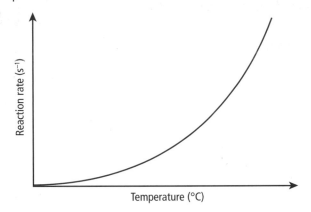

Conclusion

As the temperature increases the reaction rate increases.

Evaluation

- Factors which were kept constant were the concentrations and volumes of the different reactants. The only factor changed was the temperature of the reaction mixture.
- The beaker had to be dried each time to ensure that the concentrations of the reactants were exactly the same each time. Any water present would have made the solutions slightly more dilute.
- It can be quite difficult to be sure of exactly when the purple colour has disappeared entirely. If the reaction had been carried out at too low a temperature, such as room temperature, the colour change would have been too gradual and large errors may have been present in the measurement of the time taken.
- The beaker containing the reaction mixture was placed on a white tile to make it easier to detect the colour change.
- Safety goggles had to be worn since solutions of oxalic acid, sulphuric acid and potassium permanganate irritate the eyes. They are also harmful when swallowed.

PPA 3 – ENTHALPY OF COMBUSTION

Introduction

The enthalpy of combustion of ethanol is the energy released when one mole of ethanol is burned completely in oxygen.

$$CH_3CH_2OH(l) + 3O_2(g) \rightarrow 2CO_2(g) + 3H_2O(l)$$

Ethanol is burned in a spirit burner and the heat given out is transferred to a known volume of water in a copper can. From the temperature rise, ΔT, the heat taken in by the water can be calculated using the formula,

$$E_h = cm\Delta T$$

where

c = the specific heat capacity of water and has the value $4·18 \, kJ \, kg^{-1} \, °C^{-1}$
m = mass, in kg, of the water being heated
ΔT = the rise in temperature of the water in °C.

> **DON'T FORGET**
>
> When you use this formula m is the mass of **water** in kg and that $100 \, cm^3$ of water has a mass of $100 \, g$ which is $0·1 \, kg$. The value of c is given in the Data Booklet.

contd

PPA 3 – ENTHALPY OF COMBUSTION contd

The difference in the mass of the spirit burner at the start and at the end of the experiment is the mass of ethanol which has been burned.

This mass of ethanol has burned to give out the heat energy, E_h. From this, the amount of heat energy that would be given out when one mole of ethanol (46·0 g) burns can be calculated and this value will be equal to the enthalpy of combustion of ethanol.

Since combustion is an exothermic process, the enthalpy of combustion, ΔH, will be negative.

This experiment assumes that:

- all the heat given out by the burning ethanol goes into the water
- there is complete combustion of the ethanol
- none of the ethanol is lost by evaporation.

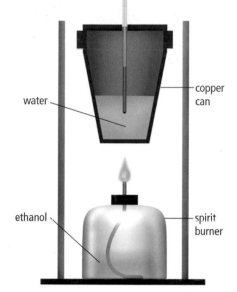

copper can

water

ethanol

spirit burner

Aim

To determine the enthalpy of combustion of ethanol.

Procedure

The five measurements made during the experiment were:

1 Initial mass of the spirit burner.
2 Volume of water in copper can.
3 Initial temperature of water.
4 Final temperature of water.
5 Final mass of the spirit burner.

Results and calculations

Initial mass of spirit burner = 110·68 g
Final mass of spirit burner = 110·42 g
Volume of water = 100 cm³
Initial temperature of water = 14°C
Final temperature of water = 26°C

From these results, you can work out that the mass of ethanol used
= 110·68 − 110·42 = 0·26 g.
$\Delta T = 26 - 14 = 12°C$
100 cm³ of water has mass 100 g = 0·10 kg

The heat taken in by the water, $E_h = cm\Delta T = 4·18 \times 0·10 \times 12 = 5·016$ kJ.
Therefore, the heat released on burning 0·26 g of ethanol = 5·016 kJ.

Ethanol has formula C_2H_5OH and its formula mass = 46, therefore one mole of ethanol = 46 g.

The heat energy released on burning one mole of ethanol $= \dfrac{5·016 \times 46}{0·26} = 887$ kJ.

Therefore, the enthalpy of combustion of ethanol = −887 kJ mol⁻¹.

Conclusion

The enthalpy of combustion of ethanol calculated from these experimental results = −887 kJ mol⁻¹.

This is much lower than the Data Booklet value of −1367 kJ mol⁻¹.

Evaluation

- Sources of error that lead to the calculated enthalpy of combustion being very different from the Data Booklet value are:

  1 Much of the heat released by the burning ethanol does not go into the water but goes to the surroundings including the air and copper can.

  2 The Data Booklet value is for **complete** combustion in oxygen. The visible carbon (soot) on the outside of the copper can indicates some incomplete combustion of the ethanol in air which is only about 20% oxygen.

  3 There is some loss of ethanol through evaporation when the cap is removed from the spirit burner.

- Ethanol is very flammable and to ensure that there is no danger of getting burned, the spirit burner was always kept in a stable position and was never refilled close to any sources of ignition.

DON'T FORGET

Combustion is an exothermic reaction and so the ΔH value must have a negative sign.

FUELS 1

PETROL

Petrol combustion

Petrol is a complex mixture of hydrocarbons ranging in size from C_5 to C_{10}. The principal source of petrol is the naphtha fraction obtained by the fractional distillation of crude oil.

In a petrol engine, a mixture of petrol and air is drawn into the cylinder. A piston pushes up into the cylinder and compresses the petrol/air mixture. At the point of maximum compression, the mixture is ignited by a spark and an explosion occurs. The thrust from the expanding product gases pushes the piston down and this mechanical energy is transmitted to the drive wheels of the vehicle enabling it to move.

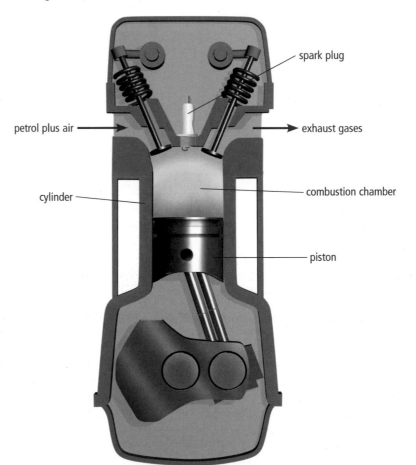

spark plug

petrol plus air

exhaust gases

cylinder

combustion chamber

piston

DON'T FORGET

'Knocking' is the sound of actual explosions in the cylinders of a car engine and is caused by the auto-ignition of certain hydrocarbons, particularly straight-chain alkanes.

When the petrol/air mixture is compressed in the cylinder, it heats up and the more it is compressed the hotter it gets. Under these conditions, some hydrocarbons, particularly straight-chain alkanes, undergo **auto-ignition**, that is they catch fire spontaneously and without the application of a spark. This premature ignition of the petrol is called 'knocking', and results in low engine performance and can cause damage to the cylinder.

Initially, the petrol produced commercially was simply the 'straight-run' naphtha fraction from crude oil – it wasn't modified in any way and was a poor quality fuel. Thereafter, it was normal practice to add lead compounds to the petrol to improve its quality. These reduced the tendency of the hydrocarbons to auto-ignite. Although the lead compounds proved to be effective 'anti-knock' agents, there were drawbacks in using them. Lead is toxic and the lead compounds issuing from car exhausts caused pollution of the environment. Leaded petrol was phased out and finally withdrawn from sale in the UK in 2000.

contd

PETROL contd

Reforming

The tendency of petrol to auto-ignite can be expressed in terms of its octane number. The higher the octane number the more efficiently it will burn. The addition of lead compounds increased the octane number of petrol but when they were phased out, chemists had to devise alternative methods of producing petrols with high octane numbers. One way of doing this is to **reform** some of the hydrocarbons in the naphtha fraction. **Reforming** is a chemical reaction in which the arrangement of atoms in a hydrocarbon molecule is altered without changing the number of carbon atoms in the molecule. The naphtha fraction contains an abundance of straight-chain alkanes and they have low octane numbers. However, when they are reformed they are converted into branched-chain alkanes, cycloalkanes and aromatic hydrocarbons which have high octane numbers. Some examples of these reforming reactions are illustrated below:

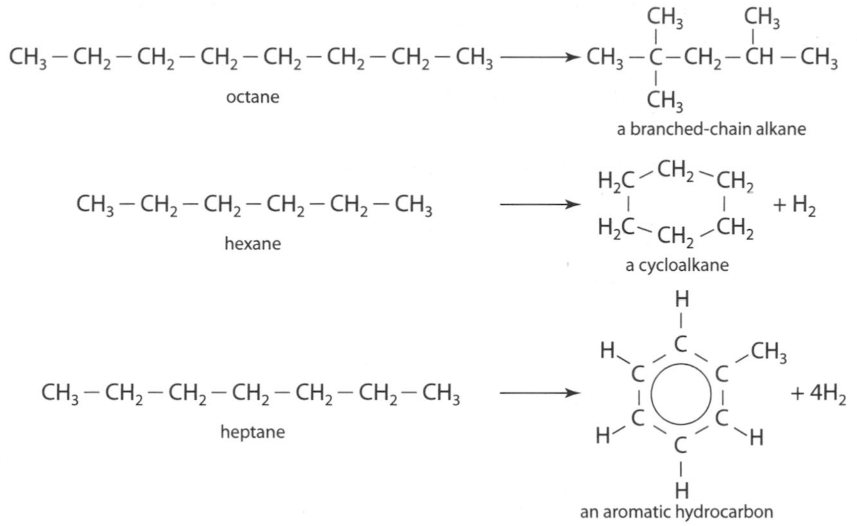

$CH_3 - CH_2 - CH_2 - CH_2 - CH_2 - CH_2 - CH_2 - CH_3 \longrightarrow$

octane

a branched-chain alkane

$CH_3 - CH_2 - CH_2 - CH_2 - CH_2 - CH_3 \longrightarrow$

hexane

a cycloalkane $+ H_2$

$CH_3 - CH_2 - CH_2 - CH_2 - CH_2 - CH_2 - CH_3 \longrightarrow$

heptane

an aromatic hydrocarbon $+ 4H_2$

> ### DON'T FORGET
>
> In reforming, the carbon skeletons of straight-chain alkanes are reorganised into branched-chain alkanes, cycloalkanes and aromatic hydrocarbons.

The products obtained in reforming naphtha are then blended with some of the straight-run naphtha fraction to give the petrol we use today. It is a high quality petrol because of the branched-chain alkanes, the cycloalkanes and the aromatic hydrocarbons that are present.

The **volatility** of the hydrocarbons is another factor that must be taken into account when making petrol. The volatility of a petrol is a measure of how easily it vaporises. In very cold weather, for example, petrol doesn't vaporise readily making the engine difficult to start. In very warm weather on the other hand, the petrol vaporises too readily and can cause a 'vapour lock' in the fuel pipe and prevent the petrol reaching the engine. To overcome these problems, petrol companies make different blends to take into account the prevailing temperatures. In winter, they add more of the smaller more volatile hydrocarbons while in summer, more of the larger less volatile hydrocarbons are added.

⚙ LET'S THINK ABOUT THIS

The **octane number** of a petrol is assigned by measuring its potential to auto-ignite compared with that of a mixture of two hydrocarbons, heptane and iso-octane, shown opposite:

Heptane, a straight-chain alkane, has a high tendency to auto-ignite and has been assigned an octane number of **0**. Iso-octane, on the other hand is a branched-chain alkane and has a low tendency to auto-ignite – it is assigned an octane number of **100**.

So, the octane number of a petrol is the percentage of iso-octane in a mixture of iso-octane and heptane that has the same tendency to auto-ignite as the petrol itself. For example, a petrol with an octane rating of 98 will have the same tendency to auto-ignite as a mixture of 98% iso-octane and 2% heptane.

$CH_3CH_2CH_2CH_2CH_2CH_2CH_3$

heptane

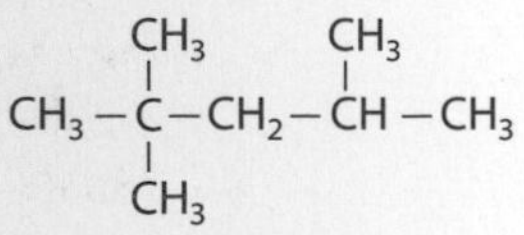

iso-octane

FUELS 2

ALTERNATIVE FUELS

Biofuels

Biofuels are fuels derived from biomass which is simply plant or animal material. The major biofuel is **ethanol** made by the fermentation of sugars. It is used extensively in Brazil where it is derived from the sucrose in sugar cane. As a motor fuel, ethanol is used both in its neat form (100% ethanol) and mixed with petrol. The mixed form is known as gasohol and consists of 20% ethanol and 80% petrol.

As a fuel, ethanol has a number of advantages:
- It is an oxygen-containing compound and so burns more efficiently than the hydrocarbons in petrol and therefore produces less of the pollutant, carbon monoxide.
- It is produced from renewable resources, such as sugar cane, and therefore is a saving on finite resources like crude oil.
- Burning ethanol derived from sugar cane makes no net contribution to the greenhouse effect. This is because the carbon dioxide absorbed from the atmosphere during the growth of the sugar cane (photosynthesis) is exactly balanced by that released into the atmosphere when the ethanol burns.

Another important biofuel is **biogas**. It is a gaseous mixture of **methane** (about 60%) and carbon dioxide and can be generated from waste biomass such as that present in sewage plants, refuse dumps and on farms. Biogas is formed by the action of bacteria on the decaying material in the waste biomass. The process is described as **anaerobic** (without air) **fermentation**.

Methanol

Methanol (CH_3OH) is another important alternative fuel. It is manufactured by passing synthesis gas – a mixture of carbon monoxide and hydrogen – over a heated catalyst:

$$CO(g) + 2H_2(g) \longrightarrow CH_3OH(l)$$

Methanol has long been used as a fuel for Formula 1 racing cars but has been suggested as a fuel for general use in cars. Advantages of using methanol as a fuel include:
- It has a high octane number.
- It burns more completely than the hydrocarbons in petrol and so produces less carbon monoxide.
- It is less volatile than petrol and is less likely to explode in a collision.

However, there are disadvantages in using methanol. For example:
- It is toxic and long-term exposure can lead to blindness and brain damage.
- It is hygroscopic, that is it absorbs water from the atmosphere. This water could cause corrosion in the engine.
- It produces less energy than the same volume of petrol. This would imply that larger fuel tanks or more frequent visits for refuelling would be required.

contd

DON'T FORGET

Ethanol is a good alternative to petrol as a fuel since it can be made from sugar cane which is a renewable source.

ALTERNATIVE FUELS contd

Hydrogen

With the diminishing supply of fossil fuels, **hydrogen** has long been considered as an attractive alternative to petrol as a fuel. On burning, it produces about three times as much energy per gram and makes no contribution to global warming since no carbon dioxide is produced, only water.

There are a number of practical problems that have to be solved before hydrogen can be considered as a realistic alternative to petrol. The most crucial of these is finding an inexpensive way of producing hydrogen. Currently, it is made by steam reforming natural gas or coal:

$$CH_4(g) + H_2O(g) \longrightarrow CO(g) + 3H_2(g)$$

$$C(s) + H_2O(g) \longrightarrow CO(g) + H_2(g)$$

While both methods are relatively inexpensive, they rely on fossil fuels as raw materials. This makes little sense if the overriding goal is to replace fossil fuels. The most obvious source of hydrogen is water which is in plentiful supply and renewable. Water can be electrolysed to generate hydrogen and solar energy could be used to drive this process. Solar energy is the logical choice since it is renewable. This method of making hydrogen is still in its early days of development and there are numerous technical problems still to be solved.

The storage and distribution of hydrogen is another problem that has to be solved. It could be stored as the gas under moderate conditions of temperature and pressure but this would be impractical because the volume occupied would be too large. In addition, storing hydrogen at high pressure or as a liquid would reduce the volume but would require special equipment and safety would be an issue. Researchers are currently looking at metals and carbon nanotubes (fullerenes) as a means of storing hydrogen. They are capable of absorbing relatively large volumes of hydrogen, for example, palladium absorbs about 1000 times its volume of hydrogen.

The term **'hydrogen economy'** has been coined to describe the overall strategy in using hydrogen as a means of storing and distributing energy.

DON'T FORGET

The use of hydrogen in the internal combustion engine would reduce the build up of carbon dioxide in the atmosphere.

LET'S THINK ABOUT THIS

Much of the **biogas** produced in the world comes from an unlikely source. It is cattle. Some of the cellulose, in the grass they eat, undergoes anaerobic fermentation resulting in them emitting biogas into the atmosphere. Biogas of course contains methane and carbon dioxide and both are greenhouse gases. In fact, methane is about 25 times more efficient at trapping heat in the atmosphere than carbon dioxide. It has been estimated that cattle and other ruminants are responsible for 18% of the greenhouse gases in the atmosphere – more than cars, planes and all other forms of transport put together. With such a significant contribution to global warming, it is not surprising that this issue is being addressed by scientists and governments with some degree of urgency.

NOMENCLATURE AND STRUCTURAL FORMULAE 1

HYDROCARBONS

Different formulae

Organic compounds can be represented using different types of **formulae**. Take **pentane** for example. It can be represented as C_5H_{12}

This is its **molecular formula** and shows the different types of atom in a molecule of pentane and the numbers of each type.

This is its **full structural formula** (drawn in the margin) and shows all the atoms and all the bonds in the pentane molecule.

These are known as **shortened structural formulae** (drawn in the margin) and are abbreviated versions of the full structural formula.

$$
\begin{array}{ccccc}
H & H & H & H & H \\
| & | & | & | & | \\
H-C-C-C-C-C-H \\
| & | & | & | & | \\
H & H & H & H & H
\end{array}
$$

$$CH_3-CH_2-CH_2-CH_2-CH_3$$
and
$$CH_3CH_2CH_2CH_2CH_3$$

Alkanes

The **alkanes** are a subset of the larger set of **hydrocarbons**. They are **saturated** since they contain no carbon-to-carbon double or triple bonds and they have the **general formula** C_nH_{2n+2}. Alkanes can be straight-chain or branched-chain.

Naming straight-chain alkanes

Straight-chain alkanes, like pentane above, are so-called because the carbon atoms are joined in one continuous chain. The first eight **straight-chain alkanes** are named in the following table along with their molecular formulae and shortened structural formulae.

DON'T FORGET

It is worthwhile committing the names of these straight-chain alkanes to memory but they can be found on page 6 of the Data Booklet should you forget them.

Name	Molecular formula	Shortened structural formula
methane	CH_4	CH_4
ethane	C_2H_6	CH_3CH_3
propane	C_3H_8	$CH_3CH_2CH_3$
butane	C_4H_{10}	$CH_3CH_2CH_2CH_3$
pentane	C_5H_{12}	$CH_3CH_2CH_2CH_2CH_3$
hexane	C_6H_{14}	$CH_3CH_2CH_2CH_2CH_2CH_3$
heptane	C_7H_{16}	$CH_3CH_2CH_2CH_2CH_2CH_2CH_3$
octane	C_8H_{18}	$CH_3CH_2CH_2CH_2CH_2CH_2CH_2CH_3$

Naming branched-chain alkanes

Naming straight-chain alkanes is relatively easy but how do we name branched-chain alkanes? Let's consider a typical **branched-chain alkane**:

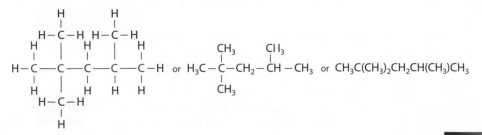

contd

HYDROCARBONS contd

We can regard this alkane as a chain of five C atoms (shaded red in the full structural formula) with three –CH₃ branches (shaded yellow) attached. The –CH₃ branch is just methane (CH₄) minus a hydrogen atom and is called a **methyl** group. Similarly, a –CH₂CH₃ branch would be called an ethyl group.

To work out the name of this alkane, we apply the following internationally agreed set of rules:

- Pick out the longest continuous carbon chain to get the parent name of the compound. The longest chain here contains five C atoms (shaded red) and so the parent name is **pentane**.
- Number the C atoms in the longest chain starting at the end nearer a branch. We hit a problem straight away because there is a branch on the second carbon atom in from the left and one on the second carbon atom in from the right. In such cases, we ignore these two branches and apply the rule again. So numbering would start at the left hand end of the chain.
- Identify the branches and arrange them in **alphabetical** order. For example, 'ethyl' comes before 'methyl' which in turn comes before 'propyl'. If there are two or more of the same branch, then this is indicated by using 'di', 'tri', 'tetra' and so on. We then identify the numbers of the C atoms in the chain to which the branches are attached and insert these into the name. In our example:

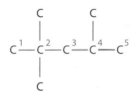

there is only one type of branch	**methyl**
but there are three of them	**trimethyl**
attached at C-2, C-2 and C-4	**2,2,4-trimethyl**

- Finally, we write '2,2,4-trimethyl' in front of the parent name and arrive at:

  2,2,4-trimethylpentane

  as the systematic name for the above branched-chain alkane.

Writing structural formulae

As well as being able to name straight-chain and branched-chain alkanes given their structural formulae, you need to be able to draw a structural formula for an alkane given its systematic name.

Take **3-ethyl-2-methylpentane**, for example. The parent name is **pentane**, so we can draw a chain of five C atoms and attach an **ethyl** group (–CH₂CH₃) at C-3 and a **methyl** group (–CH₃) at C-2. We then add H atoms to get the structure shown opposite.

The corresponding shortened structural formulae are:

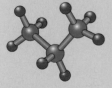   CH₃CH(CH₃)CH(CH₂CH₃)CH₂CH₃

> **DON'T FORGET**
>
> You must be able to work out the systematic names of alkanes given their full or shortened structural formulae and vice versa.

LET'S THINK ABOUT THIS

Representing the structure of a molecule in two dimensions gives a misleading picture of its shape. Take **propane**, for example. We draw its full structural formula as:

This representation suggests that the molecule is flat with bond angles of 90° but as we can see from the following picture of a propane molecule, this is far from the truth.

The four bonds around each carbon atom point towards the corners of a tetrahedron. The structural formula of propane shown on the right gives a better idea of its three-dimensional shape.

— represents a bond in the plane of the paper

⫯ represents a bond directed behind the plane of the paper

◢ represents a bond directed in front of the plane of the paper

The best way of getting a real understanding of the shapes of molecules is to build models of the molecules.

NOMENCLATURE AND STRUCTURAL FORMULAE 2

MORE HYDROCARBONS

Alkenes

The **alkenes** are another subset of the larger set of **hydrocarbons** but are distinguished from other subsets by the presence of a carbon-to-carbon double bond. The C=C double bond is the **functional group**, that is a group of atoms within a molecule which is responsible for the chemical reactions that molecule will undergo.

Since they contain a carbon-to-carbon double bond, alkenes are **unsaturated**. They have the general formula C_nH_{2n} which is the same as that for the cycloalkanes.

Naming straight-chain alkenes

Consider the two straight-chain alkenes, **A** and **B**, with molecular formula C_5H_{10}.

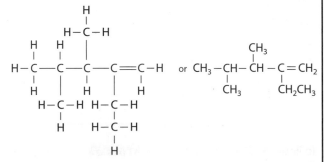

$$\begin{array}{cc} \textbf{A} & \textbf{B} \end{array}$$

$$H-\overset{1}{C}=\overset{2}{C}-\overset{3}{C}-\overset{4}{C}-\overset{5}{C}-H$$

$$CH_2=CH-CH_2-CH_2-CH_3 \qquad CH_3-CH=CH-CH_2-CH_3$$

They are both pentenes since they have five C atoms in the chain but they differ in the position of the C=C double bond. The C atoms in the chain are numbered as shown above starting at the end nearer the double bond. The C atom where the double bond starts is identified and indicated in the name.

For **A** the double bond starts at C-1 and so is called **pent-1-ene**.
For **B** the double bond starts at C-2 and is called **pent-2-ene**.

Naming branched-chain alkenes

Consider the branched-chain alkene shown opposite:

The rules for naming branched-chain alkenes are:

- Pick out the longest continuous carbon chain containing the double bond to get the parent name. In this example, the parent name would be **pentene** since the longest chain containing the double bond has five C atoms (coloured red). Notice that the longest chain in the molecule contains six C atoms but we don't use it because it doesn't contain the double bond.

- Number the C atoms in the chain starting at the end nearer the double bond. Numbering therefore starts from the right-hand end. The double bond starts at C-1 and so the parent name becomes **pent-1-ene**.

- Identify the branches (shaded yellow) and the number of the C atoms to which they are attached. There is an **ethyl** group attached at C-2 and two **methyl** groups, one attached at C-3 and the other at C-4. So, the systematic name for the above branched-chain alkene is:

2-ethyl-3,4-dimethylpent-1-ene

or $CH_3-CH-CH-C=CH_2$ with CH_3 and CH_2CH_3 branches

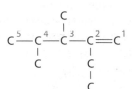

contd

DON'T FORGET

When naming alkenes, make sure you pick out the longest chain of C atoms which contains the C=C double bond.

MORE HYDROCARBONS contd

Alkynes

Alkynes are yet another subset of hydrocarbons and they have a **carbon-to-carbon triple bond** as the functional group. They are **unsaturated** and have the general formula C_nH_{2n-2}.

Alkynes are named in exactly the same way as alkenes. Consider the following alkyne:

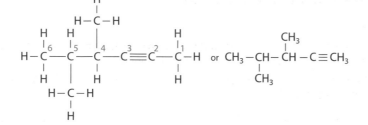

The parent name is **hexyne** since the longest carbon chain containing the triple bond has six C atoms (shaded red) in it.

Numbering starts from the right-hand end since it is nearer the triple bond. The triple bond starts at C-2 and so the parent name becomes **hex-2-yne**.

There are two branches and both are methyl groups – one is attached at C-4 and the other at C-5. So the systematic name for the above alkyne is:

4,5-dimethylhex-2-yne

> **DON'T FORGET**
>
> Given the name of an alkene or alkyne, you must be able to write its molecular formula and draw its full and shortened structural formulae.

LET'S THINK ABOUT THIS

General formulae can provide us with useful information regarding the structures of hydrocarbons. Consider the general formulae for alkanes, alkenes, cycloalkanes and alkynes.

Subset of hydrocarbons	General formula
alkanes	C_nH_{2n+2}
alkenes	C_nH_{2n}
cycloalkanes	C_nH_{2n}
alkynes	C_nH_{2n-2}

From these, we can imply that a hydrocarbon with a molecular formula containing:
- **two** fewer H atoms than its corresponding alkane will contain a C=C double bond or a ring of C atoms
- **four** fewer H atoms than its corresponding alkane may contain a C≡C triple bond. Other possibilities are that it may contain two C=C double bonds or one C=C double bond plus one ring of C atoms or even two rings of C atoms.

To illustrate these points, let's look at the following question.
A saturated hydrocarbon has the molecular formula C_5H_8.
Draw all possible structures for this hydrocarbon.

You'll notice from its molecular formula, C_5H_8, that it is four H atoms short of that for its corresponding alkane, C_5H_{12}. That implies the hydrocarbon could contain a C≡C triple bond or two C=C double bonds or one double bond and a ring of C atoms or two rings of C atoms. The other piece of crucial information given in the question is that the hydrocarbon is saturated. This narrows down the options to one, that is two rings of C atoms. So, all the possible structures are:

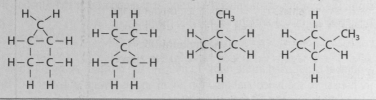

NOMENCLATURE AND STRUCTURAL FORMULAE 3

SUBSTITUTED ALKANES

Alcohols

Alcohols are a family of organic compounds in which the **hydroxyl group** or **–OH** group is the functional group.

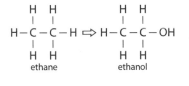
ethane ethanol

The homologous series of alcohols based on the alkanes is known as the **alkanols**. An alkanol can be regarded as a substituted alkane in which an H atom in the alkane has been replaced by an –OH group.

Naming alkanols

Straight-chain and branched-chain alkanols are named in a similar fashion to alkenes and alkynes. Consider the following alkanol:

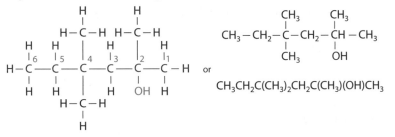

The longest carbon chain to which the –OH group is attached contains six C atoms which means the parent name is **hexanol**. Numbering starts from the right-hand end since the –OH group is nearer that end. The –OH group is attached at C-2 and so the parent name becomes **hexan-2-ol**. There are three branches and all are methyl groups – one is attached to C-2 and two are attached to C-4. The systematic name for this alkanol is therefore:

2,4,4-trimethylhexan-2-ol

Aldehydes and ketones

Aldehydes and **ketones** are two families of organic compounds which contain the **carbonyl group** ($\diagdown C=O$) as functional group.

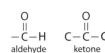
aldehyde   ketone

Although aldehydes and ketones share the same functional group there is a subtle difference in their structures. In aldehydes an **H** atom is always bonded to the carbonyl group but in ketones, the carbonyl group is always flanked by two **C** atoms.

Those aldehydes and ketones based on the alkanes are known as **alkanals** and **alkanones**, respectively.

Naming alkanals

Consider the alkanal shown opposite:

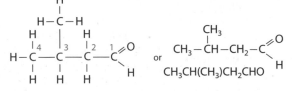

The longest carbon chain containing the carbonyl group contains four C atoms and so the parent name is **butanal**. The C atoms are numbered starting at the end nearer the functional group. A methyl group is attached at C-3 and the systematic name is:

3-methylbutanal

Notice that a number is not used in the name to indicate the position of the functional group. It is unnecessary here since the functional group in an alkanal is always at the end of the carbon chain and this end C atom is always C-1.

DON'T FORGET

The abbreviated form of the alkanal functional group is –CHO and **not** –COH.

contd

SUBSTITUTED ALKANES contd

Naming alkanones

Consider the following alkanone:

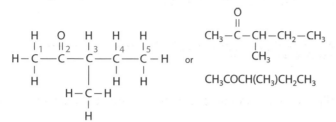

The longest chain containing the carbonyl group contains five C atoms and so the parent name is **pentanone**. Numbering starts from the left-hand end since the functional group is nearer that end. The carbonyl functional group is at C-2 and so the parent name becomes **pentan-2-one**. There is a methyl group attached at C-3 and so the systematic name is:

3-methylpentan-2-one

Carboxylic and alkanoic acids

Carboxylic acids are a family of organic compounds containing the carboxyl group

($-C{\overset{O}{\underset{OH}{\diagdown}}}$ or $-COOH$) as functional group.

The subset of carboxylic acids based on the corresponding alkanes is known as the **alkanoic acids**.

Let's consider the following alkanoic acid and name it:

$$CH_3 - CH_2 - \overset{CH_3}{\overset{|}{CH}} - CH_2 - C{\overset{O}{\underset{OH}{\diagdown}}}$$

$$CH_3CH_2CH(CH_3)CH_2COOH$$

The longest carbon chain containing the functional group has five C atoms in it and so the parent name is **pentanoic acid**. Numbering starts from the right-hand end since the functional group is nearer that end. There is a methyl group on C-3 and so the above alkanoic acid is called:

3-methylpentanoic acid

> ### DON'T FORGET
>
> In naming alkanoic acids and alkanals, a number is not required to indicate the position of the functional group since it is always at the end of the carbon chain.

LET'S THINK ABOUT THIS

The alcohol drawn below contains three hydroxyl groups. Its common name is **glycerol** but what is its systematic name?

In naming alcohols with two or more hydroxyl groups, the parent alkane name is used in full, i.e. we don't drop the 'e' off the end as we did with alcohols containing just one hydroxyl group. So, the name starts with propane- and not propan-. We use the term triol in the name to indicate that it contains three –OH groups and as usual, we must give the numbers of the C atoms at which the hydroxyl groups are attached. The systematic name of glycerol is therefore:

propane-1,2,3-triol

You will come across propane-1,2,3-triol again on page 64.

NOMENCLATURE AND STRUCTURAL FORMULAE 4

ESTERS

What are esters?

Esters are a family of organic compounds that are formed when an alkanol reacts with an alkanoic acid. For example, the ester ethyl methanoate is the organic product of the reaction between the alkanol, ethanol, and the alkanoic acid, methanoic acid:

$$
\begin{array}{cccc}
\text{H} \;\; \text{H} & \text{O} & \text{H} \;\; \text{H} & \text{O} \\
| \;\;\; | & || & | \;\;\; | & || \\
\text{H}-\text{C}-\text{C}-\text{OH} + \text{HO}-\text{C}-\text{H} \rightleftharpoons \text{H}-\text{C}-\text{C}-\text{O}-\text{C}-\text{H} + \text{H}_2\text{O} \\
| \;\;\; | & & | \;\;\; | \\
\text{H} \;\; \text{H} & & \text{H} \;\; \text{H}
\end{array}
$$

ethanol   methanoic acid   ethyl methanoate

It is the oxygen and hydrogen atoms which are coloured red in the above equation that go to form the water molecule in the reaction. Ethyl methanoate and all esters for that matter contain the following functional group:

$$
\begin{array}{c}
\text{O} \\
|| \\
\text{C}-\text{O}-\text{C}-
\end{array}
$$

It is known as the **ester group** or **ester linkage** and its presence identifies a compound as an ester. The **–oate** ending in the name also identifies a compound as an ester.

Naming esters

Esters are named after the alkanol and alkanoic acid from which they were derived. The first part of the name comes from the name of the alkanol and the second part from the name of the alkanoic acid. For example, the ester made from propan-1-ol and ethanoic acid is named propyl ethanoate and that made from ethanol and propanoic acid is called ethyl propanoate.

As well as naming an ester given the names of its parent alkanol and alkanoic acid, you must be able to name an ester given its structural formula, for example:

$$
\begin{array}{ccccccc}
\text{H} & \text{H} & \text{H} & \text{O} & \text{H} & \text{H} \\
| & | & | & || & | & | \\
\text{H}-\text{C}-\text{C}-\text{C}-\text{C}-\text{O}-\text{C}-\text{C}-\text{H} \\
| & | & | & & | & | \\
\text{H} & \text{H} & \text{H} & & \text{H} & \text{H}
\end{array}
$$

If we divide the structure into two parts through the C–O bond of the ester linkage, you will notice that the left-hand part contains the carbonyl group and **must** have been derived from the alkanoic acid. It contains four C atoms and so **butanoic acid** must have been used to make this ester. The other part must have been derived from the alkanol and since it contains two C atoms, the parent alkanol would be **ethanol**. So, the above ester is called **ethyl butanoate**.

Shortened structural formulae can also be written for esters and given these, it is possible, though more difficult, to deduce their systematic names, for example:

$$\text{CH}_3\text{CH}_2\text{CH}_2\text{COOCH}_3 \qquad \text{CH}_3\text{CH}_2\text{OOCCH}_3$$

The best way to tackle these is to draw out their full structural formula making sure we get the correct sequence of atoms in the ester group:

$$
\begin{array}{ccccccc}
\text{H} & \text{H} & \text{H} & \text{O} & \text{H} \\
| & | & | & || & | \\
\text{H}-\text{C}-\text{C}-\text{C}-\text{C}-\text{O}-\text{C}-\text{H} \\
| & | & | & & | \\
\text{H} & \text{H} & \text{H} & & \text{H}
\end{array}
\qquad
\begin{array}{ccccc}
\text{H} & \text{H} & \text{O} & \text{H} \\
| & | & || & | \\
\text{H}-\text{C}-\text{C}-\text{O}-\text{C}-\text{C}-\text{H} \\
| & | & & | \\
\text{H} & \text{H} & & \text{H}
\end{array}
$$

The ester on the left has been made from methanol and butanoic acid and is called **methyl butanoate** while that on the right is called **ethyl ethanoate** since it has been derived from ethanol and ethanoic acid.

DON'T FORGET

A compound can be identified as an ester from the functional group and the –oate ending in its name.

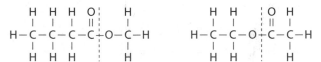
contd

ESTERS contd

Structural formulae for esters

Given the name of an ester you must be able to draw its structural formula. Take **ethyl propanoate** for example. One way of doing this is to write the equation for the reaction between its parent alkanol (ethanol) and its parent alkanoic acid (propanoic acid) like that shown on the opposite page.

Another way is to build it up from the ester linkage (shown opposite): The part on the right of the dotted line is derived from the alkanoic acid since it contains the carbonyl group. The parent alkanoic acid is propanoic acid, so we need to attach two more C atoms to the C atom on the right. One C atom has to be attached to the C atom on the left since the parent alkanol is ethanol. We then arrive at the structural formula for ethyl propanoate:

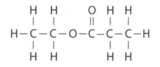

Breaking down esters

Just as an ester can be made by the reaction between an alkanol and alkanoic acid, it can also be broken back down into its parent alkanol and alkanoic acid.

Given the name of an ester, you must be able to name the products of its breakdown. We know that the first part of the name of an ester is derived from its parent alkanol and the second part from its parent alkanoic acid. So, for example, the products of the breakdown of the ester **methyl pentanoate** will be **methanol** and **pentanoic acid**.

Given the structural formula for an ester, you must also be able to identify the products of the breakdown of that ester. Consider the ester drawn below:

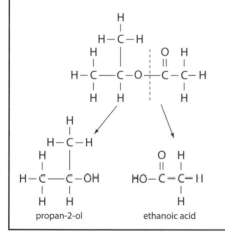

propan-2-ol   ethanoic acid

The ester will break at the point marked by the dotted red line, i.e. the O–C bond in the ester linkage. Since the right-hand part of the molecule contains the carbonyl group then it will go to form the alkanoic acid, ethanoic acid when the ester breaks down. The left-hand part will form the alkanol, propan-2-ol.

> **DON'T FORGET**
>
> Esters break down to the alkanol and alkanoic acid from which they were derived.

LET'S THINK ABOUT THIS

Esters and alkanoic acids have the same general formula of $C_nH_{2n}O_2$. This implies that the isomers of a given ester will not only include other esters but alkanoic acids as well. Take ethyl methanoate for example. It has one isomeric ester, methyl ethanoate, and one isomeric alkanoic acid, propanoic acid:

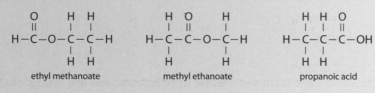

ethyl methanoate   methyl ethanoate   propanoic acid

NOMENCLATURE AND STRUCTURAL FORMULAE 5

AROMATIC HYDROCARBONS

What are aromatic hydrocarbons?

Aromatic hydrocarbons are a subset of the larger set of hydrocarbons. The simplest aromatic hydrocarbon is **benzene**. It is a colourless liquid at room temperature and has the molecular formula C_6H_6. The molecular formula shows that benzene is deficient in hydrogen atoms and suggests that it should be an unsaturated hydrocarbon. However, when benzene is added to bromine solution, rapid decolourisation does not take place. This implies that **benzene resists addition reactions** and is more stable than a typical unsaturated hydrocarbon. The reason for this will become clear when we examine the structure of benzene.

Structure of benzene

Benzene has a special ring structure and can be represented by the following structural formula:

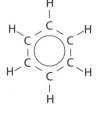

The benzene ring is flat like a regular hexagon. The carbon-to-carbon bonds in the ring are equal in length and in strength and are intermediate between a carbon-to-carbon single bond and a carbon-to-carbon double bond. Each carbon atom in the ring uses three of its four outer electrons to form covalent bonds with one hydrogen atom and two neighbouring carbon atoms. The remaining electron is free and it, along with one from each of the other five carbon atoms, goes into electron clouds which lie above and below the plane of the benzene ring:

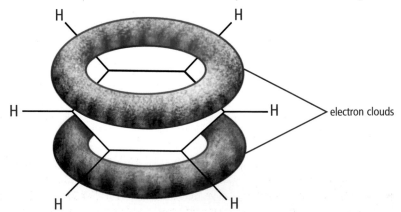

electron clouds

The six electrons that occupy the electron clouds are not tied to any one carbon atom in the ring but are shared by all six carbon atoms. They are said to be **delocalised**. In benzene's structural formula these delocalised electrons are represented by a circle drawn inside the ring. The structural formula for benzene is usually abbreviated to:

It is the delocalised electrons which give benzene its stability and makes it reluctant to undergo addition reactions.

DON'T FORGET

The stability of the benzene ring is due to the delocalised electrons.

contd

AROMATIC HYDROCARBONS contd

Other aromatic hydrocarbons

There are many other aromatic hydrocarbons, i.e. compounds like benzene, which contain six carbon atom rings stabilised by electron delocalisation. For example, if one of the hydrogen atoms in benzene is replaced by a methyl group then a hydrocarbon called **methylbenzene** (or toluene) results. It has the shortened structural formulae:

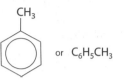 or $C_6H_5CH_3$

and molecular formula C_7H_8. Methylbenzene can be regarded as a substituted alkane. One of the hydrogen atoms in methane has been substituted by a $-C_6H_5$ or  group which is known as the **phenyl group**. So an alternative name for methylbenzene is **phenylmethane**.

Other examples of aromatic hydrocarbons include the following:

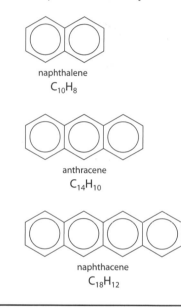

naphthalene
$C_{10}H_8$

anthracene
$C_{14}H_{10}$

naphthacene
$C_{18}H_{12}$

DON'T FORGET

The phenyl group has the formula $-C_6H_5$, or benzene minus a hydrogen atom.

LET'S THINK ABOUT THIS

Both benzene and graphite contain delocalised electrons.

Why does graphite conduct electricity yet benzene does not?

Graphite has a layer structure and each layer can be regarded as a network of fused benzene rings. The delocalised electrons extend over the whole layer and allow graphite to conduct electricity. The benzene molecule also contains delocalised electrons and this would imply that individual molecules would conduct electricity. However, a collection of benzene molecules as you would get in a beaker of the liquid does not conduct. The reason is that the delocalised electrons are confined to the individual benzene molecules and cannot 'jump' from one molecule to another.

REACTIONS OF CARBON COMPOUNDS 1

ADDITION

Addition reactions of alkenes

An alkene contains a carbon-to-carbon double bond (C=C) as the functional group and it is this reactive part of the molecule that allows an **alkene to undergo addition reactions**.

Alkenes undergo addition reactions with **hydrogen, halogens** (Cl_2, Br_2), **hydrogen halides** (HCl, HBr, HI) and **water**, for example:

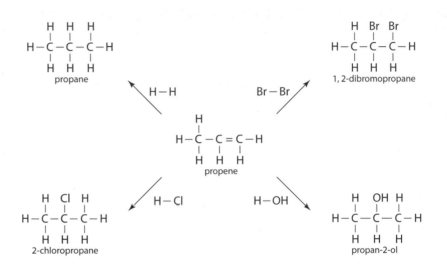

> **DON'T FORGET**
>
> Unsaturated alkenes undergo addition reactions to form saturated products.

In all four addition reactions, an **unsaturated alkene** has been converted into a **saturated compound**. The reaction of an alkene with hydrogen is also known as **hydrogenation** and the reaction with water is known as **hydration**.

Addition reactions of ethyne

Ethyne can also undergo addition reactions with hydrogen, halogens and hydrogen halides, but in a two-stage process:

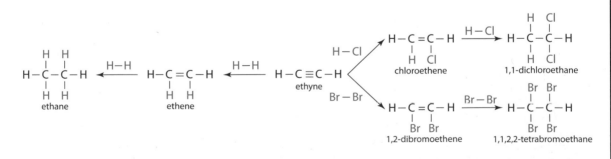

contd

ADDITION contd

Does benzene undergo addition reactions?

You will recall that benzene (C_6H_6) is an aromatic hydrocarbon and has the structural formulae shown opposite. It is the delocalised electrons (represented by the circles) which give benzene its stability and makes it reluctant to undergo addition reactions. Benzene, therefore, behaves like a saturated hydrocarbon and unlike alkenes and alkynes, it will not decolourise bromine solution.

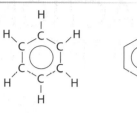

Hydration and dehydration

Ethanol used to be made in industry by fermentation but, with the exception of that made for human consumption, ethanol is now made from ethene in order to meet market demand. It involves the reaction of ethene and water in the presence of a catalyst.

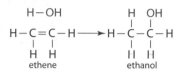

You'll recognise this as an **addition or hydration reaction**. Since a catalyst is used, it is more usually called **catalytic hydration**. With the exception of methanol, any alkanol can be made by direct catalytic hydration of its corresponding alkene.

Just as alkenes can be converted to alkanols by hydration, alkanols can be converted to alkenes by the reverse process, **dehydration**. During dehydration, the –OH group is removed along with an H atom on an adjacent carbon atom. Notice that two alkenes are formed when butan-2-ol is dehydrated. In forming but-1-ene, the –OH group is removed along with the H atom on the left and in forming but-2-ene, the H atom on the right of the –OH group is removed. With some alkanols, such as butan-1-ol, propan-2-ol and pentan-3-ol, only one alkene is formed.

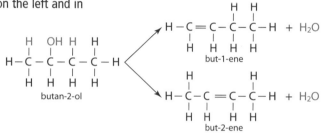

> ### DON'T FORGET
>
> Hydration is the reverse of dehydration. In hydration, water is a reactant whereas in dehydration water is a product.

LET'S THINK ABOUT THIS

When an alkene undergoes an addition reaction with a hydrogen halide or with water, **two** products may be formed. For example:

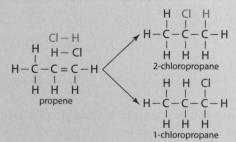

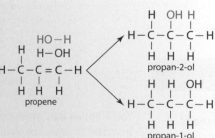

Two products will always be formed when the alkene is an **unsymmetrical** one such as propene. An alkene can be described as unsymmetrical if the groups attached to one carbon atom of the double bond are **not** identical to the groups attached to the other carbon atom.

With a symmetrical alkene, such as ethene or but-2-ene, only **one** product is formed when it reacts with a hydrogen halide or water. For example see the equation opposite:

Even if the HBr molecule added on the other way, with the Br atom becoming attached to the second carbon atom from the left, the product would still be 2-bromobutane.

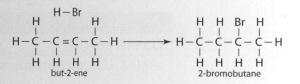

REACTIONS OF CARBON COMPOUNDS 2

OXIDATION

Oxidising alcohols

When alcohols burn in a plentiful supply of air they undergo complete combustion and are fully oxidised to carbon dioxide and water. For example:

$$CH_3CH_2OH + 3O_2 \longrightarrow 2CO_2 + 3H_2O$$

Under less severe conditions, alcohols can be partially oxidised to new organic compounds in which the carbon skeleton remains intact. The structure of the alcohol has an important bearing on the outcome of the oxidation process. There are three structural types of alcohol – **primary**, **secondary** and **tertiary**. The classification depends on the number of alkyl groups that are attached to the hydroxyl-bearing carbon atom:

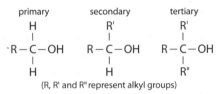

(R, R' and R" represent alkyl groups)

The primary alcohol has **one** alkyl group attached to the hydroxyl-bearing carbon atom, or no alkyl groups as is the case with methanol. The secondary has **two** alkyl groups attached and the tertiary has **three**.

Primary alcohols are oxidised to **aldehydes**, secondary alcohols are oxidised to **ketones** and tertiary alcohols resist mild oxidation. This is summarised below using the three isomeric alcohols butan-1-ol (primary), butan-2-ol (secondary) and 2-methylpropan-2-ol (tertiary).

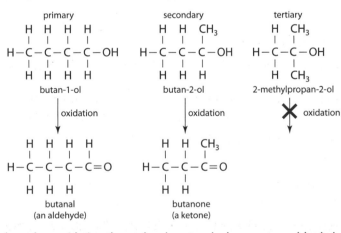

You'll notice that when oxidation does take place two hydrogen atoms (shaded red) are removed from the alcohol, one from the −OH group and one from the hydroxyl-bearing carbon atom. The tertiary alcohol has no hydrogen atom directly attached to the hydroxyl-bearing carbon atom and this is why it resists mild oxidation.

The following table shows two reagents which can be used to oxidise primary and secondary alcohols together with the results observed.

Oxidising agent	Observations
copper(II) oxide	**black** copper(II) oxide reduced to **brown** copper
acidified potassium dichromate solution	**orange** dichromate ions reduced to **blue-green** chromium(III) ions

DON'T FORGET

Primary and secondary alcohols are oxidised to aldehydes and ketones respectively while tertiary alcohols resist mild oxidation.

contd

OXIDATION contd

Oxidising carbonyl compounds

Since aldehydes and ketones both contain the carbonyl functional group, they have reactions in common but because of their structural difference there are reactions which they don't share. Aldehydes can be oxidised to **carboxylic acids** but ketones resist mild oxidation. For example:

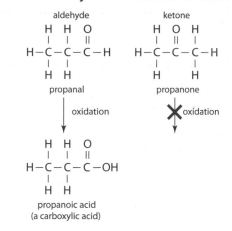

You'll notice that when an aldehyde is oxidised, an oxygen atom is inserted into the C-H bond attached to the carbonyl group. A ketone does not have a hydrogen atom attached to its carbonyl group and this accounts for the fact that it resists mild oxidation.

As well as copper(II) oxide and acidified potassium dichromate solution, Fehling's solution and Tollens' reagent can also be used to oxidise aldehydes:

Oxidising agent	Observations
Fehling's solution [alkaline solution containing $Cu^{2+}(aq)$ ions]	**blue** copper(II) ions are reduced to a **brick-red** precipitate of copper(I) oxide
Tollens' reagent [alkaline solution containing $Ag^+(aq)$ ions]	**colourless** silver(I) ions are reduced to a **grey** solid (silver metal)

Oxidation and reduction

Normally we describe oxidation and reduction in terms of a loss or gain of electrons but when applied to organic compounds:
* **oxidation** results in an **increase in the oxygen to hydrogen (O:H) ratio**
* **reduction** results in a **decrease in the oxygen to hydrogen (O:H) ratio**.

Consider the following reaction for example:

In butanone the O:H ratio is 1:8 while in butan-2-ol it is 1:10. The O:H ratio has decreased and so the above reaction is a reduction.

> **DON'T FORGET**
>
> When an organic compound is oxidised its O:H ratio increases.

LET'S THINK ABOUT THIS

Not only can **Fehling's solution and Tollens' reagent** be used to distinguish aldehydes from ketones, they can be used to **identify** an organic compound as an **aldehyde**. This is because they can only oxidise aldehydes – they are not strong enough to oxidise primary and secondary alcohols.

Another interesting fact about Tollens' reagent was its use in the manufacture of **mirrors**. A plate of glass was exposed to a mixture of Tollens' reagent and glucose. Glucose contains an aldehyde group and so reduced the silver(I) ions in the Tollens' reagent to silver metal which was deposited as a thin layer on the surface of the glass plate.

REACTIONS OF CARBON COMPOUNDS 3

MAKING AND BREAKING DOWN ESTERS

Making esters

Esters are the organic products of the reaction between carboxylic acids (alkanoic acids) and alcohols (alkanols). For example, when the carboxylic acid propanoic acid reacts with the alcohol ethanol, the ester ethyl propanoate is formed:

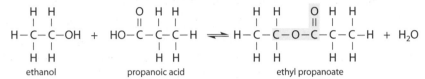

Alternatively the equation can be written as:

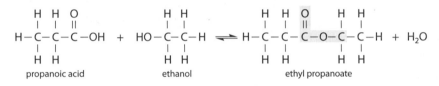

The reactant molecules join through the hydroxyl group of the alcohol and the carboxyl group of the acid with the elimination of a water molecule and the formation of the ester linkage. The hydrogen and oxygen atoms that go to form the water molecule are coloured red in the above equations and the groups of atoms which make up the ester linkages are shaded in yellow.

This reaction is described as a **condensation** reaction, that is one in which two reactant molecules join with the elimination of a small molecule which is usually water but not always. The condensation of an alcohol and a carboxylic acid to form an ester can also be referred to as an **esterification** reaction.

Condensation is a **reversible reaction** and at room temperature, it proceeds at a very slow rate. It can be speeded up by heating the reaction mixture or by adding a catalyst of **concentrated sulphuric acid**. Not only does the concentrated sulphuric acid provide the hydrogen ions needed to catalyse the reaction, it has a great affinity for water and absorbs the water that is formed in the reaction. This encourages more of the alcohol and carboxylic acid to react, thus increasing the yield of ester formed.

DON'T FORGET

The ester linkage is formed by the reaction of a hydroxyl group with a carboxyl group.

contd

MAKING AND BREAKING DOWN ESTERS contd

Breaking down esters

Since the condensation reaction that takes place in the formation of an ester is reversible, this implies that an ester can be broken back down into its parent alcohol and carboxylic acid. This involves heating the ester with water in the presence of a catalyst such as an acid or alkali. The reaction that takes place is called **hydrolysis**, for example:

butyl methanoate   butan-1-ol   methanoic acid

Let's take a closer look at what happens during the hydrolysis of butyl methanoate.

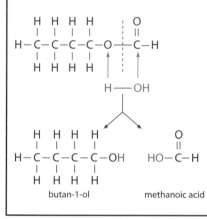

butan-1-ol   methanoic acid

The water molecule attacks the ester linkage in butyl methanoate and breaks the O–C bond (coloured red in the diagram opposite). The –OH group of the water molecule then joins with the C atom of the O–C bond to generate the carboxyl group in the product methanoic acid. The H atom left over from the water molecule joins with the O atom of the O–C bond to form the hydroxyl group of the alcohol butan-1-ol.

> **DON'T FORGET**
>
> The formation and hydrolysis of an ester is a reversible reaction.

LET'S THINK ABOUT THIS

Condensation and dehydration reactions are often confused since water is a product in both of them. Let's compare them:

- a typical **condensation reaction**
- a typical **dehydration reaction**

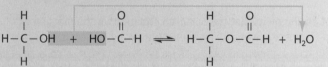 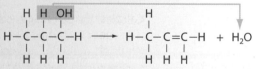

You'll notice in the condensation reaction that the water molecule is removed from **two** reactant molecules whereas in the dehydration reaction the water molecule is taken from just the **one** reactant molecule.

Hydrolysis and hydration is another pair of reactions which are often confused since in both cases water is a reactant. Let's now compare these:

- a typical **hydrolysis reaction**
- a typical **hydration reaction**

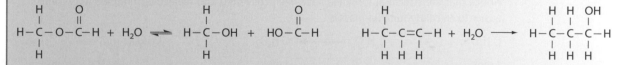

In the hydrolysis reaction, the water breaks down the ester into **two** product molecules, while in the hydration reaction the water adds on to the alkene to form only **one** product molecule.

Condensation and hydrolysis reactions are the reverse of each other and so too are dehydration and hydration reactions.

REACTIONS OF CARBON COMPOUNDS 4

PERCENTAGE YIELDS

Types of yield

The **yield** of a chemical reaction is the amount of product obtained in the reaction. There are two types of yield:

- the **theoretical yield** – this is the maximum amount of product that could be obtained if there was 100% conversion of reactants into products
- the **actual yield** – this is the amount of product that is obtained in practice.

The actual yield is usually less than the theoretical yield. There can be several reasons for this and these include:

- The reaction may be reversible which means conversion of reactants into products will never be 100%.
- Side reactions may occur in addition to the main reaction and the formation of the side products will inevitably reduce the yield of the main product.
- Even when 100% conversion is achieved, some of the product is likely to be lost when it is separated from the reaction mixture and purified.
- The initial reactants may not be 100% pure.

The actual yield of product can be expressed as a percentage of the theoretical yield,

$$\text{Percentage yield} = \frac{\text{actual yield}}{\text{theoretical yield}} \times 100$$

> **DON'T FORGET**
>
> The actual yield is the amount of product obtained in practice while the theoretical yield is the amount of product obtained had there been 100% conversion of reactants into products.

Percentage yield calculations

Consider the following examples:

Example 1

When excess acetic anhydride was added to 14·4 g of salicylic acid, 6·26 g of aspirin was obtained.

Calculate the percentage yield of aspirin given that the balanced equation for the reaction is:

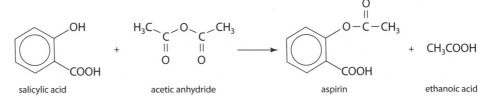

salicylic acid   acetic anhydride   aspirin   ethanoic acid

Answer

We must first calculate the theoretical yield of aspirin assuming 100% conversion of reactants into products. We can do this in the normal way using the balanced equation for the reaction. Notice that the structural formulae have been converted into molecular formulae to make it easier to calculate formula masses.

$$C_7H_6O_3 \quad + \quad C_4H_6O_3 \longrightarrow C_9H_8O_4 \quad + \quad C_2H_4O_2$$

1 mol   1 mol

138 g $\longleftarrow\longrightarrow$ 180 g

14·4 g $\longleftarrow\longrightarrow$ $180 \times \dfrac{14 \cdot 4}{138}$

$$= 18 \cdot 78\,g$$

contd

PERCENTAGE YIELDS contd

The theoretical yield of aspirin is 18·78 g and the actual yield is 6·26 g. So, by substituting these values in the percentage yield expression we obtain:

$$\text{percentage yield} = \frac{6\cdot26}{18\cdot78} \times 100 = 33\cdot3\%$$

Example 2

Ammonia is made by the Haber process:

$$N_2(g) + 3H_2(g) \longrightarrow 2NH_3(g)$$

Under test conditions, the percentage yield of ammonia in the Haber process was found to be 14%. Calculate the mass of hydrogen that would be needed to react with excess nitrogen to give an actual yield of ammonia of 150 kg.

Answer

We have been given the percentage yield and the actual yield of ammonia and so by rearranging the percentage yield expression we can work out the theoretical yield of ammonia, that is the yield for 100% conversion.

$$\text{percentage yield} = \frac{\text{actual yield}}{\text{theoretical yield}} \times 100 \text{ rearranges to theoretical yield} = \frac{\text{actual yield}}{\text{percentage yield}} \times 100$$

$$\text{theoretical yield} = \frac{150}{14} \times 100 = 1071 \text{ kg}$$

Now we can use the balanced equation to calculate the mass of hydrogen that would be needed.

$$N_2 \quad + \quad 3H_2 \longrightarrow 2NH_3$$

3 mol   2 mol

6 g   34 g

$$6 \times \frac{1071 \times 10^3}{34} \longleftrightarrow 1071 \times 10^3 \text{ g}$$

$$= 189 \times 10^3 \text{ g}$$

$$= 189 \text{ kg}$$

LET'S THINK ABOUT THIS

Many substances are made not by a single reaction step but by a sequence of steps. So, how will the overall % yield of the desired product be affected by the % yields in the individual steps? To help answer this question, suppose compound **C** was made in two steps starting with compound **A**,

A $\xrightarrow{\text{step 1}}$ B $\xrightarrow{\text{step 2}}$ C
1 mol   1 mol   1 mol

Let's further suppose that the % yield in step 1 was 70% and that in step 2 was 60%.

It's tempting to say that the overall % yield of **C** would be an average of the two, that is 65%, but this would be naive. If we started with 1·00 mol of **A** then the yield of **B** would be 0·70 mol since the % yield in step 1 is 70%. In step 2, only 60% of the 0·70 mol of **B** would be converted into **C**. So, the actual yield of **C** would be 0·42 mol compared with a theoretical yield of 1·00 mol. This implies that the

$$\text{overall percentage yield} = \frac{0\cdot42}{1\cdot00} \times 100 = 42\%$$

This represents a dramatic fall from the 70% and 60% yields in the individual steps. A situation like this would not be tolerated in the chemical industry and explains why chemists are always seeking direct routes with high percentage yields when synthesising chemicals.

USES OF CARBON COMPOUNDS

COMPETING DEMANDS

There are competing demands for the fossil fuels, coal, oil and natural gas since the organic compounds derived from them can be used in two ways. Their main use is as **fuels** but they are also used in the manufacture of **consumer products** such as plastics, detergents, dyes, pesticides, medicines and so on.

Uses of esters

DON'T FORGET

Esters are used in fragrances, flavourings and solvents.

Esters have strong sweet smells and tastes which are often floral or fruity. This allows them to be used in the perfume industry as **fragrances** and in the food industry as **flavourings**.

Esters tend to be volatile, that is they have relatively low boiling points. They are often used as **solvents** where rapid evaporation of the solvent is required. For example, ethyl ethanoate is found in glues and pentyl ethanoate is used in nail varnish.

Uses of carboxylic acids

Carboxylic acids are used in a wide variety of ways. Vinegar, for example, is a dilute solution of ethanoic acid.

Ethanoic acid is also used in the manufacture of ethenyl ethanoate which can undergo addition polymerisation to form poly(ethenyl ethanoate) which is a major component of vinyl emulsion paints.

ethenyl ethanoate

The dicarboxylic acid hexanedioic acid is one of the monomers used in the manufacture of nylon.

hexanedioic acid

Uses of halogenoalkanes

Halogenoalkanes can be regarded as alkanes in which one or more of the hydrogen atoms in the alkane has been substituted by a halogen atom. Many of them are unreactive, stable, non-flammable and have a low toxicity. These properties made them highly desirable as

- **anaesthetics** – the first halogenoalkane to find use as an anaesthetic was chloroform ($CHCl_3$). It was found to damage the liver and was replaced by 2-bromo-2-chloro-1,1,1-trifluoroethane commonly called halothane.

halothane

- **aerosol propellants** – an aerosol can contained a pressurised mixture of a halogenoalkane and a substance such as a deodorant, an insecticide, a herbicide and so on. When the valve on the can was opened, the pressure inside dropped and this caused the halogenoalkane to vaporise. As it escaped into the atmosphere, it carried along with it the other substance in the mixture.
- **refrigerants** – halogenoalkanes were used in fridges and air-conditioning units because they could be easily liquefied under pressure.
- **solvents** – many halogenoalkanes can dissolve non-polar substances such as oil and grease and this is why they were used as solvents in dry-cleaning.

DON'T FORGET

The depletion of the ozone layer is believed to have been caused by the extensive use of certain CFCs.

However, in the 1980s it was discovered that certain halogenoalkanes known as chlorofluorocarbons, or CFCs for short, were responsible for the destruction of the ozone layer. The ozone layer is important to us and our environment because the ozone molecules (O_3) in it absorb much of the damaging ultraviolet radiation coming from the sun. CFCs were replaced by HCFCs (hydrochlorofluorocarbons) and HFCs (hydrofluorocarbons) and although they don't damage the ozone layer, they are greenhouse gases and contribute to global warming.

contd

COMPETING DEMANDS contd

Uses of aromatic compounds

Benzene and its related compounds are important in the manufacture of consumer products. In the table below, some examples of aromatic compounds are shown together with the consumer products made from them.

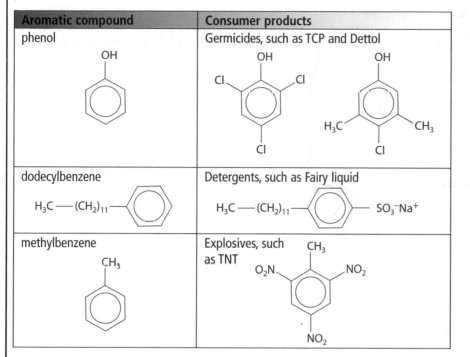

Aromatic compound	Consumer products
phenol	Germicides, such as TCP and Dettol
dodecylbenzene	Detergents, such as Fairy liquid
methylbenzene	Explosives, such as TNT

While you are not expected to remember these structures, you should be able to recognise them as aromatic compounds.

LET'S THINK ABOUT THIS

How do the chlorofluorocarbons (CFCs) damage the ozone layer?

In order to illustrate the process, let's consider the CFC, dichlorodifluoromethane (CF_2Cl_2). Initially, the ultraviolet radiation from the sun breaks a C–Cl bond in the CF_2Cl_2 molecule to form a chlorine atom:

$$CF_2Cl_2 \longrightarrow CF_2Cl + Cl$$

The chlorine atom is highly reactive and goes on to attack an ozone molecule:

$$Cl + O_3 \longrightarrow OCl + O_2$$

Like a Cl atom, OCl is highly reactive and attacks another ozone molecule:

$$OCl + O_3 \longrightarrow 2O_2 + Cl$$

The Cl atom regenerated in the last step goes on to react with other ozone molecules and so the process continues. In this way, enormous 'holes' are created in the ozone layer as the ozone molecules break down into oxygen molecules.

POLYMERS 1

EARLY PLASTICS AND FIBRES

Addition polymers

The monomer units required to make addition polymers normally contain a carbon-to-carbon double bond, that is they are unsaturated. The simplest of these is ethene and it is of major importance in the plastics industry since not only does it undergo addition polymerisation to form poly(ethene), other monomers can be derived from it to make a wide variety of addition polymers. Propene is another important monomer in the plastics industry and is used to make poly(propene). Ethene and propene are made by cracking the ethane and propane obtained from the gas or naphtha fractions from crude oil:

$$CH_3CH_3 \longrightarrow CH_2=CH_2 + H_2 \qquad CH_3CH_2CH_3 \longrightarrow CH_3CH=CH_2 + H_2$$

DON'T FORGET

Condensation polymers are made from monomers with at least two functional groups per molecule.

Condensation polymers

Each of the monomer units needed to make condensation polymers must contain at least two functional groups. With such an arrangement, each monomer is able to condense with two others and so on. Only in this way can long-chain polymer molecules be constructed.

Examples of condensation polymers include:

- **Polyesters**
  Polyesters are made from a **diol**, that is an alcohol containing two hydroxyl groups, and a **diacid**, a carboxylic acid containing two carboxyl groups. Consider the polyester made by polymerising a mixture of the diol, ethane-1,2-diol, and the diacid, benzene-1,4-dicarboxylic acid:

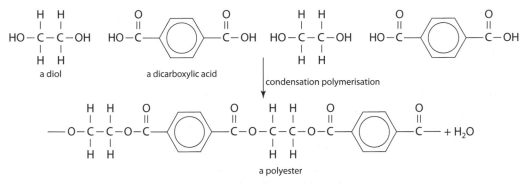

The resulting polyester sample consists of a tangled arrangement of long-chain polymer molecules with weak van der Waals' bonds between them. It is a thermoplastic and so can be melted without it decomposing. In industry, the molten polyester is forced through tiny holes and cooled to form thin **fibres**. On stretching these fibres, the polymer chains uncoil and line up alongside each other. The polymer molecules are now packed more closely together and as a result, the intermolecular bonds are stronger. This coupled with the very strong covalent bonds within the molecules make the fibres strong. In fibre form the above polyester is marketed under the trade name, Terylene. Polyesters can also be manufactured in the form of **resins**. The resins differ in structure from fibres in that they contain covalently bonded cross links between adjacent chains giving a rigid and strong three-dimensional structure. Some polyester resins are mixed with glass fibres to form glass reinforced plastic which is used in making canoes, dinghies and car-body panels.

- **Polyamides**
  Amines are another important family of organic compounds and they contain the **amino group** (–NH$_2$) as functional group. They condense with carboxylic acids to form amides, for example:

The group of atoms coloured red is known as the **amide link**.

contd

EARLY PLASTICS AND FIBRES contd

Just as diols condense with diacids to form polyesters, diamines condense with diacids to form **polyamides**. The first polyamide was formed by polymerising a mixture of the monomers 1,6-diaminohexane and hexanedioic acid:

This particular polyamide is called nylon-6,6 and is just one of a variety of nylons. Although polyamides are thermoplastic polymers, they have relatively high melting points and it is the presence of the amide links that accounts for this.

$$H-\overset{\overset{\displaystyle H}{|}}{N}-(CH_2)_6-\overset{\overset{\displaystyle H}{|}}{N}-H \quad HO-\overset{\overset{\displaystyle O}{||}}{C}-(CH_2)_4-\overset{\overset{\displaystyle O}{||}}{C}-OH \quad H-\overset{\overset{\displaystyle H}{|}}{N}-(CH_2)_6-\overset{\overset{\displaystyle H}{|}}{N}-H \quad HO-\overset{\overset{\displaystyle O}{||}}{C}-(CH_2)_4-\overset{\overset{\displaystyle O}{||}}{C}-OH$$

a diamine   a diacid

↓ condensation polymerisation

$$-\overset{\overset{\displaystyle H}{|}}{N}-(CH_2)_6-\overset{\overset{\displaystyle H}{|}}{N}-\overset{\overset{\displaystyle O}{||}}{C}-(CH_2)_4-\overset{\overset{\displaystyle O}{||}}{C}-\overset{\overset{\displaystyle H}{|}}{N}-(CH_2)_6-\overset{\overset{\displaystyle H}{|}}{N}-\overset{\overset{\displaystyle O}{||}}{C}-(CH_2)_4-\overset{\overset{\displaystyle O}{||}}{C}- \quad + \quad H_2O$$

a polyamide

Both the N–H and C=O bonds in the amide link are polar since nitrogen has a greater electronegativity than hydrogen, and oxygen has a greater electronegativity than carbon. As a result, each atom in the amide link has a slight charge and this allows hydrogen bonds to be set up between adjacent polyamide chains:

The hydrogen bonds are coloured red in the diagram. Since hydrogen bonds are stronger than van der Waals' forces, polyamides have a higher melting point and are considerably stronger than polyesters. It is this increased strength that allows some nylons to be used as engineering plastics, ie to make machine parts and so on.

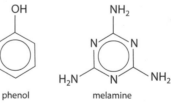

Thermosetting plastics

Another important group of condensation polymers are the methanal-based polymers. Bakelite and melamine resin are just two examples of this type of polymer and they are made by condensing methanal with phenol and melamine, respectively. The resulting polymers have a cross-linked three-dimensional network structure, ie they are thermosetting plastics.

phenol   melamine

The methanal needed to make this group of polymers can be manufactured in a series of reactions starting with methane (from natural gas) or carbon in the form of coal. In the first stage of the process, methane (CH_4) or coal (mainly carbon) is **steam reformed** to form a mixture of carbon monoxide and hydrogen which is known as **synthesis gas**:

$$CH_4(g) + H_2O(g) \longrightarrow CO(g) + 3H_2(g) \quad C(s) + H_2O(g) \longrightarrow CO(g) + H_2(g)$$

The **synthesis gas** is then converted into methanol which is finally oxidised to methanal:

$$CO(g) + 2H_2(g) \longrightarrow CH_3OH(l) \text{ followed by } CH_3OH(l) \longrightarrow HCHO(l)$$

DON'T FORGET

Synthesis gas is a mixture of carbon monoxide and hydrogen and can be obtained by steam reforming methane or coal.

LET'S THINK ABOUT THIS

What is the significance of the numbers in nylon-6,6? They simply correspond to the numbers of carbon atoms in the diamine ($H_2NCH_2CH_2CH_2CH_2CH_2CH_2NH_2$) and the diacid ($HOOCCH_2CH_2CH_2CH_2COOH$) used to make it. Nylon-6 is a similar polymer to nylon-6,6 and its name implies that only one monomer is used in its manufacture. This, in turn, implies that the monomer must contain both an amino group and a carboxyl group:

$$H-\overset{\overset{\displaystyle H}{|}}{\underset{\underset{\displaystyle H}{|}}{N}}-\overset{\overset{\displaystyle H}{|}}{\underset{\underset{\displaystyle H}{|}}{C}}-\overset{\overset{\displaystyle H}{|}}{\underset{\underset{\displaystyle H}{|}}{C}}-\overset{\overset{\displaystyle H}{|}}{\underset{\underset{\displaystyle H}{|}}{C}}-\overset{\overset{\displaystyle H}{|}}{\underset{\underset{\displaystyle H}{|}}{C}}-\overset{\overset{\displaystyle H}{|}}{\underset{\underset{\displaystyle H}{|}}{C}}-\overset{\overset{\displaystyle O}{||}}{C}-OH$$

POLYMERS 2

RECENT DEVELOPMENTS

Kevlar

Kevlar is a synthetic fibre of exceptional **strength**. Part of its structure is illustrated below:

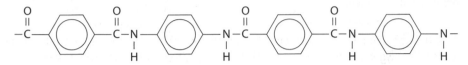

DON'T FORGET

Kevlar is an aromatic polyamide which is extremely strong because of the way in which the rigid linear molecules are packed together.

You will notice that Kevlar is an aromatic polyamide and it is made by the condensation reaction between a diamine and a diacid. Its strength is due to the way in which the polymer chains pack together. The rigid, rod-like chains are themselves very strong and this strength is increased when they pack closely together into flat sheets with hydrogen bonds operating between adjacent chains. Not only is it very strong, Kevlar is also resistant to heat and to abrasion and doesn't react with other chemicals. With this unique combination of properties, it finds use in many specialist ways. For example, it has replaced steel in reinforced tyres and is used in making brake pads, clutch linings and cables. It is also used in bullet-proof vests and in the protective clothing worn by fire-fighters and motorcyclists.

Poly(ethenol)

Unlike the vast majority of synthetic polymers, poly(ethenol) is **soluble** in water and it is this property which has allowed it to be exploited commercially. Poly(ethenol) is made by treating the addition polymer, poly(ethenyl ethanoate), with methanol in a process known as 'ester exchange':

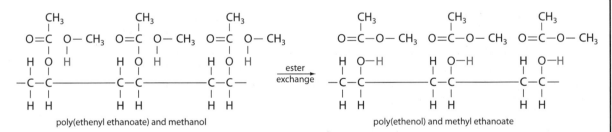

poly(ethenyl ethanoate) and methanol   poly(ethenol) and methyl ethanoate

In the 'ester exchange' process, notice that the acid part (coloured red) of the ester side chains of the poly(ethenyl ethanoate) is removed and exchanged for the hydrogen atom (coloured green) in the –OH group of methanol. By carefully controlling the reaction conditions, a variety of poly(ethenol)s can be made depending on the number of ester side chains in the poly(ethenyl ethanoate) that are exchanged. It is the polar –OH groups in the poly(ethenol)s that account for their water solubility and the degree of solubility depends on the percentage of –OH groups present. It is this variable water solubility which makes the poly(ethenol)s so useful. Hospital laundry bags, for example, are made from a poly(ethenol) which is soluble in hot water but insoluble in cold. Dirty linen can be safely contained in the bag until it is placed in the washing machine, at which point the bag dissolves in the hot water and releases the laundry into the wash. Poly(ethenol)s are also used by surgeons for internal stitching. The type used depends on how long the stitches have to remain in place before they dissolve.

contd

RECENT DEVELOPMENTS contd

Poly(ethyne)

The most striking property of poly(ethyne) is its ability to **conduct electricity** when certain chemicals called 'dopants' are introduced into its structure. It is an addition polymer made from ethyne monomer units:

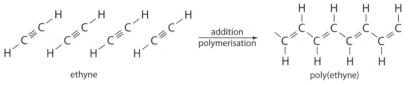

ethyne → (addition polymerisation) → poly(ethyne)

No practical application has yet been found for doped poly(ethyne) because it is unstable in air – it reacts with oxygen and water vapour and, as a result, its electrical conductivity rapidly falls.

Poly(vinyl carbazole)

Poly(vinyl carbazole) is an addition polymer made from the unsaturated monomer vinyl carbazole (shown opposite). When poly(vinyl carbazole) is doped it becomes **photoconductive** which means it conducts electricity on exposure to light. Its photoconductivity has made it an ideal component of the coating on drums in photocopiers and laser printers.

vinyl carbazole

Biopol

Biopol is a natural polymer synthesised by a bacterium. The bacterium uses it as an energy store just like plants use starch and animals use fats. Industrial chemists exploited this polymer-making bacterium and used it to produce Biopol on a commercial scale. Part of the structure of Biopol is illustrated opposite:

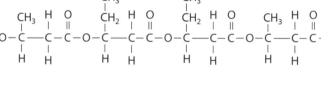

You will notice that Biopol is a condensation polymer made from two different monomer units and since it contains ester linkages, it can also be described as a polyester. Being natural, Biopol is **biodegradable** which gives it a major advantage over traditional polymers. It was used as a packaging material for cosmetics and motor oils up until 1999 when production ceased because of its high production costs and a shift in attitude among manufacturers away from biodegradable polymers to ones that can be recycled.

> **DON'T FORGET**
>
> Biopol is a natural polyester and as such is biodegradable.

Photodegradable low density polythene

One way of dealing with the growing problem of polymer waste disposal is to take existing polymers and redesign them in such a way that they become degradable. Low density polythene is one such polymer. It has been redesigned to **degrade on exposure to ultraviolet light**. To achieve this, carbonyl groups were introduced into the backbone of polythene during its manufacture.

The carbonyl groups have been picked out in red. On exposure to sunlight, they absorb and trap the ultraviolet radiation. The energy is used to break carbon-to-carbon bonds in the vicinity of the carbonyl groups. As a result, the polymer chains are broken into smaller fragments which biodegrade more rapidly.

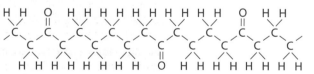

Photodegradable low density polythene is currently being used to make agricultural sheeting.

LET'S THINK ABOUT THIS

Why does doped poly(ethyne) conduct electricity? If you look at the structure of poly(ethyne) you will notice that each carbon atom must use three of its four outer electrons to bond with three other atoms. The remaining electron together with one from each of the other carbon atoms in the chain, are delocalised. This means that individual poly(ethyne) molecules will be able to conduct but a collection of them will not. However, the dopant molecules 'bridge' the gaps between the chains and allow the delocalised electrons to move from one molecule to another. In this way, the whole sample of poly(ethyne) conducts and not just individual molecules within it.

NATURAL PRODUCTS 1

FATS AND OILS

The structure of fats and oils

Natural fats and oils can be classified according to their origin as:
- **animal**, for example, beef fat, butter fat, pork fat (lard)
- **vegetable**, for example, olive oil, sunflower oil, linseed oil, rapeseed oil
- **marine**, for example, cod-liver oil, sardine oil, whale oil.

They are an essential part of our diet and supply the body with energy. Since the percentage of oxygen in fats and oils is less than that in carbohydrates, they are a more concentrated source of energy.

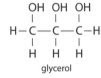
glycerol

Fats and oils are examples of **esters** and are formed by condensation reactions between the alcohol, **glycerol**, and carboxylic acids known as **fatty acids**.

Glycerol is described as a trihydric alcohol because it contains three hydroxyl groups and it has the systematic name propane-1,2,3-triol.

Fatty acids are straight-chain carboxylic acids containing even numbers of carbon atoms ranging from C_4 to C_{24} but more commonly C_{16} and C_{18}. They can be saturated like stearic acid, for example, or unsaturated like oleic acid:

$$CH_3CH_2CH_2CH_2CH_2CH_2CH_2CH_2CH_2CH_2CH_2CH_2CH_2CH_2CH_2CH_2CH_2COOH$$
stearic acid (saturated)

$$CH_3CH_2CH_2CH_2CH_2CH_2CH_2CH=CHCH_2CH_2CH_2CH_2CH_2CH_2CH_2CH_2COOH$$
oleic acid (unsaturated)

Since a glycerol molecule contains three hydroxyl groups, it will condense with three fatty acid molecules to form the esters present in fats and oils:

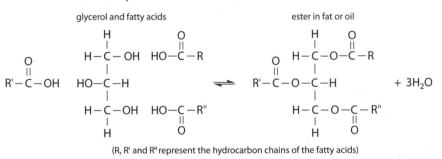

glycerol and fatty acids   ester in fat or oil

(R, R' and R'' represent the hydrocarbon chains of the fatty acids)

An ester formed from glycerol is called a glyceride and so the esters present in fats and oils are referred to as triglycerides. Any fat or oil contains a mixture of triglycerides, some in which all the fatty acid groups are the same, others in which they are different.

Properties of fats and oils

In general, oils decolourise bromine solution to a much greater extent than fats indicating that the degree of unsaturation in oils is greater. This difference in the degree of unsaturation accounts for the fact that oils tend to be liquids at room temperature while fats are solids, ie oils have lower melting points than fats. Since oil molecules contain more carbon-to-carbon double bonds than fat molecules, their shapes are considerably different. Oil molecules are less compact and can't pack together as closely as fat molecules. This implies that the van der Waals' forces between oil molecules are weaker and therefore easier to break than those between fat molecules thus explaining why oils have lower melting points.

DON'T FORGET

The lower melting points of oils compared to those of fats is related to the higher degree of unsaturation in oil molecules.

contd

FATS AND OILS contd

Reactions of fats and oils

When oils are heated with hydrogen in the presence of a nickel catalyst, the hydrogen molecules add on across some of the carbon-to-carbon double bonds in the oil molecules. As a result, the unsaturated nature of the oil molecules is partially removed and in effect, unsaturated oils are converted into saturated fats. This process is known as **hardening** since it converts soft liquid oils into harder solid fats and the reaction taking place can be described as hydrogenation or addition. Margarine is made in this way from oils such corn oil and soya bean oil.

Just as a simple ester can undergo **hydrolysis** (reaction with water) to form an alcohol and carboxylic acid, fats and oils can also be hydrolysed to produce glycerol and fatty acids:

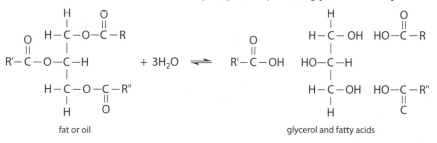

fat or oil   glycerol and fatty acids

Notice that for every mole of fat or oil that is hydrolysed, one mole of glycerol and three moles of fatty acids are formed.

This hydrolysis reaction forms the basis of **soap making**. In the manufacture of soaps, fats or oils are boiled with sodium (or potassium) hydroxide solution. The alkali first catalyses the hydrolysis reaction and then neutralises the fatty acids produced to form their sodium (or potassium) salts:

fatty acid from fat or oil   soap

It is these sodium (or potassium) salts of fatty acids that are soaps.

> **DON'T FORGET**
>
> Soaps are the sodium and potassium salts of fatty acids and are produced by the hydrolysis of fats and oils.

LET'S THINK ABOUT THIS

Another use for vegetable oils, particularly rapeseed oil, is in the production of **biodiesel**, an alternative to the conventional diesel fuel. To produce it, the rapeseed oil is treated with methanol in the presence of a catalyst and the following reaction takes place:

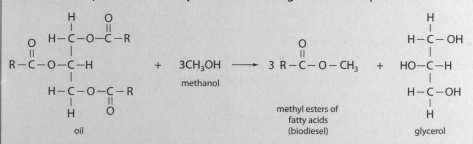

oil   methanol   methyl esters of fatty acids (biodiesel)   glycerol

The glycerol part of the ester molecule in the oil has been replaced by methanol to give methyl esters of fatty acids, i.e. biodiesel. This reaction is an example of ester exchange, the same as that involved in making poly(ethenol) (see page 62).

Biodiesel production, however, is not without its detractors. It is argued that in a world of food shortages, it makes little sense to use up land in growing the crops needed to make biodiesel.

NATURAL PRODUCTS 2

PROTEINS

Organic nitrogen compounds

Nitrogen is found in a wide variety of organic compounds but the most important of these are **proteins**. Proteins occur in all living organisms. In the human body for example, they are found in muscle, hair, skin, blood and so on, and make up about one sixth of our bodyweight. Nearly all the nitrogen in proteins is derived originally from nitrate ions present in the soil. Only plants can use this simple form of fixed nitrogen in making protein; animals can not. Animals therefore must eat plant proteins or other animal proteins and reconstruct them into the proteins they need.

Amino acids

Proteins are **condensation polymers** made from **amino acid monomer units**. Although about twenty different amino acids are used in synthesising proteins, they can all be represented by the general structure shown opposite:

They all have in common an amino group (–NH$_2$) and a carboxyl group (–COOH) attached to the same carbon atom but they differ in the R group.

For example, when R=H, the amino acid is called glycine (or aminoethanoic acid) and when R=CH$_3$, it is alanine (or 2-aminopropanoic acid).

Polypeptides and proteins

Since amino acids contain both a basic amino group and an acidic carboxyl group, they are able to condense with each other to form larger molecules. When two amino acids condense together, a dipeptide is formed. For example:

The group of atoms coloured red in the dipeptide is the same as an amide link and when it is present in a peptide it is more commonly referred to as a **peptide link**. When three amino acids condense together, a tripeptide is formed and when a large number link up, a **polypeptide** is produced:

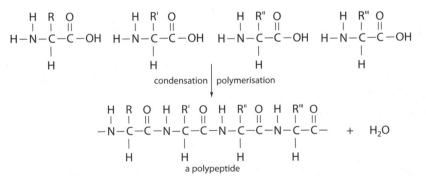

Proteins are polypeptides and range in length from about 40 amino acid units to over 4000.

As mentioned earlier, about twenty different amino acids are used to make proteins. So with twenty different choices available for each amino acid unit in a polypeptide chain, it is not surprising that there are a huge number of different proteins.

DON'T FORGET

Proteins are condensation polymers made up of many amino acid monomer units linked together.

contd

PROTEINS contd

Hydrolysis of proteins

In the laboratory, proteins can be broken down into their amino acid monomer units by heating them with dilute acid. The proteins are hydrolysed when they react with water and the process is catalysed by the hydrogen ions of the acid:

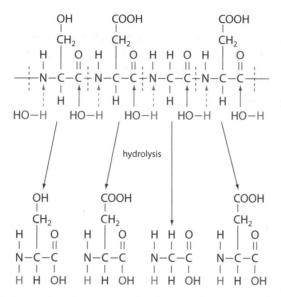

During the hydrolysis reaction the water molecules attack and break the C–N bonds (coloured red in the above diagram) of the peptide links and the individual amino acid molecules are generated as shown.

While plants are able to synthesise all the amino acids they need to make proteins, animals can not. Those amino acids that we, as animals, can't synthesise are known as **essential amino acids**. We obtain them by hydrolysing the plant and/or animal proteins in the food we eat. During digestion of the food, the proteins are hydrolysed and the amino acids that are formed pass into the bloodstream and are carried to various sites in the body where they are reassembled into the specific proteins we need.

Fibrous and globular proteins

Protein chains contain polar peptide links along their length.

These peptide links can hydrogen bond with others in the same chain to produce a **spiral** or with others in adjacent chains to form a so-called **pleated sheet**. The spirals or pleated sheets are then assembled into more complicated structures known as **fibrous** or **globular** proteins. **Fibrous proteins** are long and thin and are the major structural materials of animal tissue. Examples include keratin (hair, nails, wool), collagen (tendons, cartilage) and muscle proteins like actin and myosin. In globular proteins, the spiral polypeptide chains are folded into more compact units. **Globular proteins** are involved in the maintenance and regulation of life processes and include enzymes, hormones, like insulin, and the oxygen-carrying protein in the blood, haemoglobin.

DON'T FORGET

Given the structure of a section of a protein, you must be able to draw the structural formulae of the amino acids obtained when it is hydrolysed.

LET'S THINK ABOUT THIS

Each protein has a unique primary structure, that is the sequence in which the amino acid units are bonded to one another in the protein chain. Insulin was the first protein to be sequenced and this was achieved by Frederick Sanger in 1955. He hydrolysed the insulin and then used chromatography to identify the amino acids and the order in which they were linked in the chain. This achievement earned him his first Nobel Prize in 1958, and in 1977, a postage stamp was issued commemorating his work.

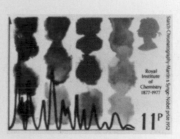

NATURAL PRODUCTS 3

MORE ON PROTEINS

Enzymes

Enzymes are globular proteins and they catalyse the chemical reactions in living organisms. Like inorganic catalysts, they do this by lowering the activation energies of the reactions but, unlike inorganic catalysts, enzymes are highly specific. Each enzyme will only catalyse one specific reaction or type of reaction. Enzymes are also classified as homogeneous catalysts since they are in the same physical state as the reactants.

The catalytic activity of enzymes is affected by changes in temperature and by changes in pH. The following graphs show how the rate of starch hydrolysis varies with temperature and with pH in the presence of the enzyme, amylase:

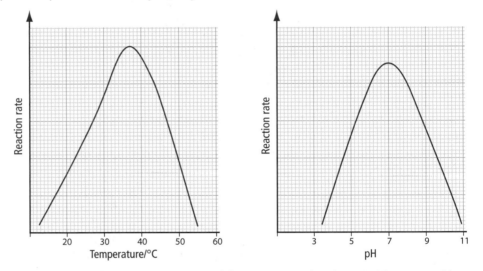

The graph on the left shows that the rate of this enzyme-catalysed reaction increases with temperature up to about 37°C but then decreases and falls to zero above 50°C when the enzyme is destroyed. 37°C is said to be the optimum temperature of amylase, that is the temperature at which it operates most efficiently. All enzymes in the human body share an optimum temperature of 37°C. This is not surprising since it corresponds to body temperature. Other enzymes have different optimum temperatures. For example, the enzymes present in the bacteria which inhabit hot springs have much higher optimum temperatures and zymase, which catalyses the fermentation of glucose, has an optimum temperature of about 27°C.

The graph on the right shows that the enzyme amylase has an optimum pH of 7. Although enzymes in the body share the same optimum temperature, they don't necessarily have the same optimum pH. The latter depends on the prevailing pH conditions at the location of the enzyme. Consider, for example, pepsin and trypsin which catalyse protein hydrolysis. Pepsin (present in the stomach) has an optimum pH of 2 while trypsin (present in the small intestine) has its maximum activity at pH 8.

DON'T FORGET

Enzymes are highly specific and are most efficient within a narrow range of temperature and pH.

contd

MORE ON PROTEINS contd

Mechanism of enzyme catalysis

Enzymes are globular proteins and roughly spherical in shape. All of them have 'clefts' or 'slots', known as **active sites**, on their surfaces. It is believed that the molecule on which the reaction is performed, namely the **substrate molecule**, binds on to the active site. The reaction on the substrate molecule takes place and the product molecules vacate the active site leaving it free to accommodate another substrate molecule. The active site has a particular shape and can only accept a substrate molecule which exactly matches that shape. This is the reason why enzymes are highly specific in their catalytic activity.

This mechanism is illustrated below for the hydrolysis of starch by the enzyme amylase. Starch is a condensation polymer made from the monosaccharide, glucose, and on hydrolysis it is broken down into maltose, a disaccharide made up of two glucose units.

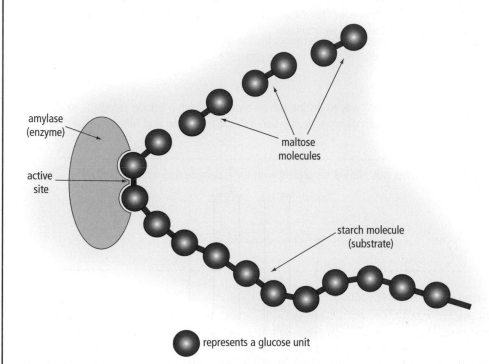

amylase (enzyme)

active site

maltose molecules

starch molecule (substrate)

● represents a glucose unit

We have already learned that an enzyme functions best at a particular temperature and pH and if conditions deviate from these optimum values, the catalytic activity of the enzyme is lowered significantly. Changes in temperature and pH alter the shape of the active site. As a result, the enzyme has difficulty in 'recognising' its substrate molecule because the shape of its active site no longer matches precisely that of its substrate molecule.

In general, if enzymes are heated to temperatures in excess of about 55°C or are subjected to extremes of pH, then their catalytic activity is lost completely. The reason for this is that the structure of the enzyme is irretrievably changed and the active site is destroyed. When this happens the enzyme is said to be **denatured**.

DON'T FORGET

When an enzyme is denatured, its molecular shape is changed, its active site is destroyed and it loses its catalytic activity.

LET'S THINK ABOUT THIS

Enzymes are highly efficient catalysts and are able to speed up reactions by enormous amounts. Typically, enzyme-catalysed rates are 10^7 to 10^{14} times faster than uncatalysed rates. To get a better handle on this, let's consider an enzyme-catalysed reaction with a reaction time of one second. If the enzyme increased the rate by a factor of 10^7 then the reaction time for the uncatalysed reaction would be 116 days. Similarly, for an enzyme which increased the rate by 10^{14}, the uncatalysed reaction time would be about three million years. These simple calculations demonstrate the vital role that enzymes play – without enzymes biochemical reactions would be far too slow to sustain life.

PPA 1–3

PPA 1 – OXIDATION

Introduction

Aldehydes and ketones both contain the carbonyl functional group, $\supset C = O$

In aldehydes, the carbonyl group has a hydrogen atom attached whereas in ketones, it is flanked by two carbon atoms as shown below:

It is this difference in structure that accounts for the fact that aldehydes can undergo mild oxidation to form carboxylic acids but ketones can not.

Aim

The aim of this experiment was to distinguish an aldehyde from a ketone using the oxidising agents, acidified potassium dichromate solution, Fehling's solution and Tollens' reagent.

Procedure

Each of the reagents was heated separately with the aldehyde and with the ketone in a hot water bath.

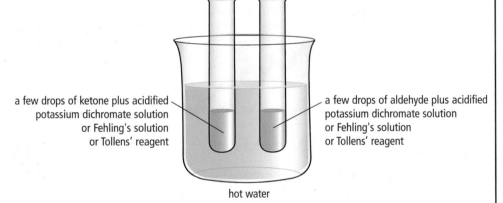

a few drops of ketone plus acidified potassium dichromate solution or Fehling's solution or Tollens' reagent

a few drops of aldehyde plus acidified potassium dichromate solution or Fehling's solution or Tollens' reagent

hot water

At the start,

- the acidified potassium dichromate solution has an orange colour
- Fehling's solution (an alkaline solution containing copper(II) ions) has a blue colour
- Tollens' reagent (an alkaline solution containing silver(I) ions) is colourless.

Results

With the aldehyde,

- the orange-coloured acidified potassium dichromate solution turned blue-green
- the blue-coloured Fehling's solution changed to give a brick-red precipitate
- the colourless Tollens' reagent changed to give a shiny grey solid (silver) which was deposited on the walls of the test tube.

With the ketone, no changes took place.

Conclusion

Acidified potassium dichromate solution or Fehling's solution or Tollens' reagent can be used to distinguish an aldehyde from a ketone.

> **DON'T FORGET**
>
> Aldehydes are oxidised to carboxylic acids but ketones resist mild oxidation.

> **DON'T FORGET**
>
> You need to know the colour changes that take place when an aldehyde is heated with acidified potassium dichromate solution, with Fehling's solution and with Tollens' reagent.

contd

PPA 1 – OXIDATION contd

Evaluation

- The hot water bath had to be set up and all flames extinguished before the reaction mixtures were prepared. This is to reduce the risk of fire since the aldehyde and ketone are both flammable.

- Any one of the above oxidising agents can be used to distinguish between an aldehyde and a ketone but only Fehling's solution and Tollens' reagent can be used to identify an organic compound as an aldehyde.

- Gloves had to be worn when working with Tollens' reagent and the aldehyde and ketone because of the damage they can do to the skin.

- Great care had to be exercised with all the chemicals because of their hazardous nature.

PPA 2 – MAKING ESTERS

Introduction

One way of preparing esters is to condense an alcohol with a carboxylic acid:

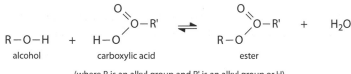

alcohol   carboxylic acid   ester

(where R is an alkyl group and R' is an alkyl group or H)

The reaction is slow at room temperature but the rate can be increased by heating the reaction mixture and using a catalyst of concentrated sulphuric acid. Not only does the concentrated sulphuric acid provide the hydrogen ions needed to catalyse the reaction, it has a great affinity for water and absorbs the water that is formed in the reaction. This encourages more of the alcohol and carboxylic acid to react, thus increasing the yield of ester formed.

Aim

The aim of this experiment was to prepare an ester and to identify some of the characteristic properties of esters.

Procedure

The alcohol, carboxylic acid and a few drops of concentrated sulphuric acid were added to a test tube and a wet paper towel was wrapped round the neck and secured with a rubber band. After inserting a cotton wool plug in the mouth of the test tube it was then placed in a hot water bath as shown opposite.

After about 10 minutes the reaction mixture was slowly poured into some sodium hydrogencarbonate solution and cautiously smelled. Any sign of the ester separating from the aqueous mixture was noted.

Results

The ester formed a separate layer on top of the aqueous layer and it had a pleasant smell.

Conclusion

The following equation shows the formation of a typical ester:

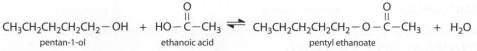

pentan-1-ol   ethanoic acid   pentyl ethanoate

The ester was insoluble in water, less dense than water and had a characteristic pleasant smell.

> **DON'T FORGET**
>
> The reaction between an alcohol and a carboxylic acid is an example of a condensation reaction.

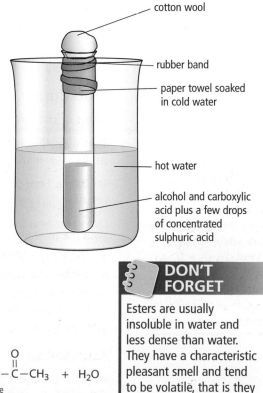

cotton wool

rubber band

paper towel soaked in cold water

hot water

alcohol and carboxylic acid plus a few drops of concentrated sulphuric acid

> **DON'T FORGET**
>
> Esters are usually insoluble in water and less dense than water. They have a characteristic pleasant smell and tend to be volatile, that is they evaporate readily.

contd

PPA 2 – MAKING ESTERS contd

Evaluation

- The hot water bath had to be set up and all flames extinguished before the reaction mixture was prepared. This is to reduce the risk of fire since the alcohol, carboxylic acid and the ester product are all flammable.
- Gloves had to be worn when working with concentrated sulphuric acid since it is highly corrosive.
- The reaction mixture was heated and a catalyst of concentrated sulphuric acid was used to increase the reaction rate.
- The purpose of the wet paper towel round the neck of the test tube was to act as a condenser. This prevented the escape of the volatile ester, alcohol and carboxylic acid.
- The plug of cotton wool in the neck of the test tube was to provide protection should the reaction mixture 'spurt' during heating.
- The purpose of the sodium hydrogencarbonate solution was to neutralise the sulphuric acid and any unreacted carboxylic acid.
- Vigorous effervescence (bubbles of carbon dioxide released) took place when the reaction mixture was poured into sodium hydrogencarbonate solution. The pouring was done slowly and carefully to prevent the reaction mixture 'frothing' over or 'spurting' out.

PPA 3 – FACTORS AFFECTING ENZYME ACTIVITY

Introduction

The enzyme catalase catalyses the decomposition of hydrogen peroxide:

$$2H_2O_2(aq) \longrightarrow 2H_2O(l) + O_2(g)$$

In this PPA, potato is used as the source of catalase and the rate of the enzyme catalysed reaction is found by counting the bubbles of oxygen released from the reaction mixture during a set period of time.

Aim

The aim of this experiment was to investigate the effect of pH or temperature changes on enzyme activity.

Procedure

Effect of changing pH

$5\,cm^3$ of a pH 7 buffer solution and three potato discs were added to the side-arm test tube and left for three minutes.

On adding $1\,cm^3$ of hydrogen peroxide to the test tube, a timer was immediately started and the test tube stoppered.

During the next three minute period, the number of bubbles given off were counted and recorded.

The experiment was repeated using buffer solutions of pH 1, 4, 10 and 13.

> **DON'T FORGET**
>
> Enzymes are globular proteins and are highly specific – each enzyme will only catalyse one particular reaction or type of reaction.

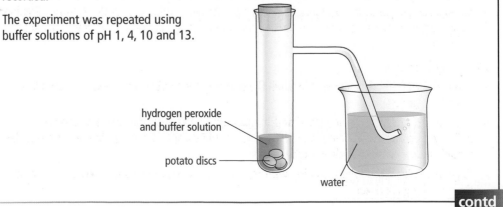

hydrogen peroxide and buffer solution

potato discs

water

contd

PPA 3 – FACTORS AFFECTING ENZYME ACTIVITY contd

Effect of changing temperature

5 cm³ of water and three potato discs were added to the side-arm test tube which was placed in a water bath. The mixture was left to stand until the temperature remained constant and that steady temperature was measured and recorded.

On adding 1 cm³ of hydrogen peroxide to the test tube, a timer was immediately started and the test tube stoppered.

During the next three minute period, the number of bubbles given off were counted and recorded.

The experiment was repeated another four times and before each 'run', the water in the bath was heated to approximately 30°C, 40°C, 50°C and 60°C.

water

hydrogen peroxide and buffer solution

water bath

potato discs

Results

Some typical results from both investigations are shown in the tables opposite.

Effect of changing pH		Effect of changing temperature	
pH	Number of gas bubbles released in 3 minute period	Temperature (°C)	Number of gas bubbles released in 3 minute period
1	3	19	8
4	9	31	15
7	20	39	28
10	15	52	5
13	4	63	0

Conclusions

Effect of changing pH

Catalase activity increases as pH increases to about pH7 and then decreases **or** the optimum pH of catalase is approximately 7.

Effect of changing temperature

Catalase activity increases as temperature increases to about 39°C and then decreases **or** the optimum temperature of catalase is about 39°C.

> **DON'T FORGET**
>
> Enzymes are most efficient within a narrow range of pH and temperature.

Evaluation

pH investigation

- The potato disc/buffer solution mixture was left to stand for about three minutes before adding the hydrogen peroxide solution in order to give the catalase time to adjust to the pH of the buffer.

- In order to obtain valid results, only the pH of the solution was varied. Other factors, like the temperature of the solution, number of potato discs, volume of buffer solution, volume of hydrogen peroxide solution and time period for counting bubbles of gas, had to be kept the same.

Temperature investigation

- The potato disc/water mixture was left to stand in the water bath until the temperature remained steady to ensure that the catalase had reached that temperature.

- In order to obtain valid results, only the temperature of the solution was varied. Other factors, like the pH of the solution, number of potato discs, volume of buffer solution, volume of hydrogen peroxide solution and time period for counting bubbles of gas, had to be kept the same.

Both investigations

- The activity of the enzyme was measured by counting the bubbles of oxygen released in a fixed time period.

- The stopper had to be tightly fitted and the connection between the side-arm and the delivery tube had to be air-tight, otherwise gas would escape at these points and no bubbles would emerge from the delivery tube.

- Care had to be taken when working with the hydrogen peroxide solution because it irritates the eyes, lungs and skin.

THE CHEMICAL INDUSTRY

THE UK CHEMICAL INDUSTRY

The UK chemical industry is a major contributor to the quality of our everyday lives. Were it not for the chemical industry, thousands of consumer products which we take for granted would not exist. For example, there would be no plastics, synthetic fibres, detergents, pharmaceuticals, fuels like petrol and diesel, paints, toiletries, and so on. The chemical industry also plays an important role in our national economy. It is the only major sector of the UK manufacturing industry to make a positive contribution to the country's balance of trade with the rest of the world. It also provides direct employment for about 250 000 people and indirectly supports several hundred thousand additional jobs.

The chemical industry in the UK was first established in Scotland and the North of England close to sources of its raw materials and in locations with good transport links. Even now, with much improved communication networks, new chemical plants are often built in these same areas. One reason for this is the ready access to a pool of skilled labour that has built up over the years.

THE MANUFACTURING PROCESS

The major **raw materials** from which all chemicals are ultimately derived are the fossil fuels, water, air, metal ores and minerals. They provide the **feedstocks**, that is the reactants, which go into a chemical process. For example, the raw materials in the Haber process to make ammonia are natural gas (methane), water (as steam) and air. The natural gas and water provide the feedstock, hydrogen, and the air provides nitrogen. In preparing feedstocks from raw materials it is absolutely crucial that all impurities be removed. Otherwise, catalysts could be poisoned, yields could be lowered and the final product could be contaminated.

Manufacture of chemicals from the pure feedstock can be carried out either as a **batch process** or as a **continuous process**. In batch processing, the feedstock is fed into a vessel and allowed to react. When the reaction is complete the products are removed and separated. Batch processing is normally used when relatively small quantities of chemicals, such as pharmaceuticals, dyes, pesticides and so on, are to be made. In continuous processing, the feedstock is fed into one end of the plant and products drawn out the other end in a continuous flow. It is used in manufacturing very large quantities of chemicals, such as ethene, ammonia, sulphuric acid and polythene. There are advantages and disadvantages of both batch and continuous processing. For example, continuous processing:

- is less flexible since the plant is dedicated to the manufacture of one particular chemical,

- requires a smaller workforce since it is more easily automated,

- involves building a plant for the particular product which becomes much more expensive,

- is more time efficient since shut-downs (times when no product is being made) can be years apart. In batch processing shut-downs can happen on a daily basis as the reaction vessel is emptied of product, cleaned and then recharged with feedstock.

> **DON'T FORGET**
>
> Feedstocks are the reactants in a chemical process. Be careful not to confuse 'feedstocks' and 'raw materials'. Hydrogen, for example, is a feedstock in the Haber process but it can not be classified as a raw material.

ASPECTS OF CHEMICAL MANUFACTURING

There are a number of important aspects of chemical manufacturing, some of which are considered below.

- **Costs** – there are three types of cost: capital, fixed and variable. The main capital costs are incurred in building the plant and its infrastructure. Fixed costs include rates, salaries, repaying loans and so on, and are incurred no matter whether a plant is working at full

contd

ASPECTS OF CHEMICAL MANUFACTURING contd

capacity or not. Variable costs include the cost of raw materials and the cost of distributing the product. Since there is a huge investment of capital to build plants and relatively few people employed in operating them, the UK chemical industry can be described as capital intensive rather than labour intensive.

- **Energy** – the efficient use of energy is an important consideration in most chemical processes because of its high cost. Many chemical reactions are exothermic and the heat energy released can be conserved by lagging pipes and using heat exchangers. Energy from exothermic steps in the process can be used to supply energy for endothermic steps. Using catalysts which operate at lower temperatures is another way of reducing energy costs.

- **Operating conditions of temperature and pressure** – these are chosen to maximise economic efficiency. For example, one step in the manufacture of sulphuric acid is carried out at a pressure slightly above atmospheric pressure despite the fact that higher pressures would increase the yield. The increase in yield, however, is only marginal and so in terms of cost, the use of high pressures cannot be justified.

- **Choice of synthetic route** – a chemical can usually be made in a number of different ways and many factors influence which synthetic route is chosen to manufacture the chemical. Ethanoic acid, for example, can be made either by direct oxidation of naphtha or by the catalysed reaction between methanol and carbon monoxide. In both, the operating conditions of temperature and pressure are roughly the same. The first process has the advantage that no expensive catalyst is needed but the yield of ethanoic acid is very low. The low yield implies that considerable quantities of by-products are formed and these could have presented marketing problems. In the event, markets were available for these by-products which made the manufacture of ethanoic acid from naphtha an economically viable process. The major advantage of the second process is the high yield of ethanoic acid (over 99%) and the flexibility of feedstock production – methanol can be made from any of the fossil fuels. The major disadvantage is the high cost of the catalyst. The methanol/carbon monoxide route to ethanoic acid is now preferred and all new plants use this process. Existing naphtha oxidation plants however are unlikely to be replaced until maintenance costs become too high or the bottom falls out of the market for the by-products. So the choice of a particular synthetic route is influenced by factors which include cost, availability and suitability of feedstock, yield of product, opportunities for the recycling of reactants and marketability of any by-products.

- **Safety and environmental issues** – these are major considerations in the chemical industry and subject to rigorous legislation. Computer systems, for example, are used to monitor the process and will automatically shut down any part of the plant that isn't functioning properly. Air quality in the vicinity of the plant is also monitored and warning systems are in place to alert local residents should any abnormal emissions be detected.

DON'T FORGET

The UK chemical industry is, by and large, capital rather than labour intensive.

MANUFACTURING A NEW PRODUCT

The manufacture of a completely new product begins in the research lab where a method of synthesising the compound is devised. The next stage is to carry out a pilot study in which larger quantities of the compound are made. At this stage equipment similar to that used in final production is employed rather than the apparatus used in the small-scale experiments carried out in the research lab. Finally, the process is scaled up and the compound goes into full production. At each of these stages, the information obtained is reviewed and any necessary modifications to the process are made.

LET'S THINK ABOUT THIS

The term 'chemical' conjures up 'toxic substance' in the minds of many members of the general public. They fail to appreciate that everything in this world including them is made up of chemicals, the vast majority of which are completely harmless. Furthermore, any accident in the chemical industry always attracts bad press and this fuels the negative image the public has of the industry. Although the only acceptable accident rate is zero, it is worth pointing out that working in the chemical industry is significantly less hazardous than crossing the street. While the industry itself expends considerable effort to improve its image, we as chemists must not abrogate our role in this process.

HESS'S LAW

CALCULATING ENTHALPY CHANGES USING HESS'S LAW

What is Hess's Law?

Hess's Law states that the enthalpy change for a chemical reaction is independent of the route taken.

This statement follows on from the law of Conservation of Energy – energy cannot be created or destroyed.

Imagine a reaction which can take place by two different routes.

B can be made directly from A (route 1) or via C (route 2).

According to Hess's Law, $\Delta H_1 = \Delta H_2 + \Delta H_3$

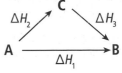

Manipulating chemical equations

Calculations are usually done by first writing out the balanced chemical equation for the enthalpy change you are trying to calculate.

Other chemical equations are then manipulated to give the desired equation. These equations can be reversed or multiplied. Whatever operation is carried out on the equation, then the same operation has to be carried out on the value of the enthalpy change.

For example, the equation for the combustion of hydrogen is:

$$H_2(g) + \tfrac{1}{2}O_2(g) \longrightarrow H_2O(l) \qquad\qquad \Delta H = -286\,kJ\,mol^{-1}$$

If in a calculation this equation has to be doubled, then the value for the enthalpy change will also have to be doubled to $-572\,kJ$.

If the equation has to be reversed to:

$$H_2O(l) \longrightarrow H_2(g) + \tfrac{1}{2}O_2(g) \text{ then the value of } \Delta H \text{ becomes } +286\,kJ.$$

Calculations using Hess's Law

The equation which represents the formation of ethane from carbon and hydrogen is:

$$2C(s) + 3H_2(g) \longrightarrow C_2H_6(g) \;\; \Delta H = ?$$

In practice, carbon and hydrogen do not react to form ethane and so ΔH for the above reaction cannot be determined experimentally. However, a calculation based on Hess's Law can be used to determine the value of the enthalpy change.

The enthalpies of combustion of carbon, hydrogen and ethane have been determined experimentally and are given in the Data Booklet, page 9.

These can be used to calculate ΔH for $2C(s) + 3H_2(g) \longrightarrow C_2H_6(g)$ as shown below.

Step 1

Write the balanced chemical equations for the combustion of carbon, hydrogen and ethane and note the enthalpy of combustion values from the Data Booklet.

[1] $C(s) \quad + \quad O_2(g) \quad \longrightarrow \quad CO_2(g) \qquad\qquad \Delta H = -394\,kJ\,mol^{-1}$

[2] $H_2(g) \quad + \quad \tfrac{1}{2}O_2(g) \quad \longrightarrow \quad H_2O(l) \qquad\qquad \Delta H = -286\,kJ\,mol^{-1}$

[3] $C_2H_6(g) \quad + \quad 3\tfrac{1}{2}O_2(g) \quad \longrightarrow \quad 2CO_2(g) + 3H_2O(l) \qquad \Delta H = -1560\,kJ\,mol^{-1}$

DON'T FORGET

Whatever operation is carried out to the equation, then that same operation must be carried out to the enthalpy change value. If the equation is reversed, then the sign of the enthalpy change value must be changed.

contd

CALCULATING ENTHALPY CHANGES USING HESS'S LAW contd

Step 2

Write a balanced equation representing the desired enthalpy change.

$$2C(s) + 3H_2(g) \longrightarrow C_2H_6(g) \qquad \Delta H = ?$$

This is the target equation and we can use Hess's Law to get to this equation from the three equations written on the opposite page in Step 1.

Step 3

This involves using equations [1], [2] and [3] to get the target equation.

In the target equation there are two carbons on the left side and in equation [1] there is one carbon on the left side. Equation [1] therefore has to be multiplied by 2 and so does the enthalpy change value. Similarly equation [2] has to be multiplied by 3 since the target equation has $3H_2$ on the left side and equation [2] has only one H_2 on the left side. The enthalpy change value will also have to be multiplied by 3.

The target equation has $C_2H_6(g)$ on the right side whereas equation [3] has $C_2H_6(g)$ on the left side. Equation [3] has to be reversed and the sign of the enthalpy change value changed.

Putting these changes into place:

[1] × 2: $\qquad 2C(s) + 2O_2(g) \longrightarrow 2CO_2(g) \qquad \Delta H = -384 \times 2 = -788 \text{ kJ}$

[2] × 3: $\qquad 3H_2(g) + 1\frac{1}{2}O_2(g) \longrightarrow 3H_2O(l) \qquad \Delta H = -286 \times 3 = -858 \text{ kJ}$

Reverse [3]: $\quad 2CO_2(g) + 3H_2O(l) \longrightarrow C_2H_6(g) + 3\frac{1}{2}O_2(g) \qquad \Delta H = +1560 \text{ kJ}$

Adding up these three equations gives:

$$2C(s) + 3H_2(g) + 3\tfrac{1}{2}O_2(g) + 2CO_2(g) + 3H_2O(l) \longrightarrow 2CO_2(g) + 3H_2O(l) + C_2H_6(g) + 3\tfrac{1}{2}O_2(g)$$

Removing $3\frac{1}{2}O_2(g) + 2CO_2(g) + 3H_2O(l)$ which appear on both sides of the equation gives:

$$2C(s) + 3H_2(g) \longrightarrow C_2H_6(g) \text{ which is the target equation.}$$

Since adding these three equations gives the target equation, then ΔH for this reaction will be the sum of the enthalpy change values for the three equations.
So $\Delta H = -788 - 858 + 1560 = -86 \text{ kJ mol}^{-1}$

This shows how Hess's Law can be used to calculate the enthalpy change for a reaction which cannot be carried out in practice.

LET'S THINK ABOUT THIS

If there is an oxygen in the target equation, obviously it is not possible to write an equation for the combustion of oxygen. However, oxygen will be present in the equations for other substances burning and when the equations are written and then manipulated correctly the desired number of moles of oxygen molecules in the target equation will remain.

For example, if you have to calculate the enthalpy of formation of ethanol from enthalpy of combustion data in the Data Booklet, you would use the equations:

[1] $C(s) + O_2(g) \longrightarrow CO_2(g) \qquad \Delta H = -394 \text{ kJ mol}^{-1}$

[2] $H_2(g) + \frac{1}{2}O_2(g) \longrightarrow H_2O(l) \qquad \Delta H = -286 \text{ kJ mol}^{-1}$

[3] $C_2H_5OH(l) + 3O_2(g) \longrightarrow 2CO_2(g) + 3H_2O(l) \qquad \Delta H = -1367 \text{ kJ mol}^{-1}$

The target equation for the enthalpy of formation of ethanol from its elements is:

$$2C(s) + 3H_2(g) + \tfrac{1}{2}O_2 \longrightarrow C_2H_5OH(g) \qquad \Delta H = ?$$

You should find that if you multiply equation [1] by 2 and equation [2] by 3 and reverse equation [3] and then add these up, you will get the target equation including the $\frac{1}{2}O_2(g)$ on the left side as required. Carrying out the same operations on the enthalpy change values will give the enthalpy of formation of ethanol. (Try this for yourself, you should get the answer $\Delta H = -279 \text{ kJ mol}^{-1}$.)

CHEMICAL REACTIONS

EQUILIBRIUM 1

THE CONCEPT OF DYNAMIC EQUILIBRIUM

Reversible reactions

In a chemical reaction, reactants change into products and some reactions that you have come across so far, such as boiling an egg or wool growing on a sheep, for example, most definitely cannot be reversed.

However, some reactions such as the conversion of hydrated cobalt(II) chloride into anhydrous cobalt(II) chloride is an example of a **reversible reaction**, that is the reaction can go backwards or forwards. This is shown in the two equations below.

$$CoCl_2.6H_2O \xrightarrow{\text{heat}} CoCl_2 + 6H_2O$$

pink   blue

When pink hydrated cobalt(II) chloride is heated, the water is driven off and blue anhydrous cobalt(II) chloride is the product.

$$CoCl_2 + 6H_2O \longrightarrow CoCl_2.6H_2O$$

blue   pink

When water is added to the blue anhydrous cobalt(II) chloride it changes back to pink hydrated cobalt(II) chloride.

Another example of a reversible reaction is the thermal decomposition of ammonium chloride. This is carried out by heating ammonium chloride in a test tube to produce ammonia and hydrogen chloride. The ammonia and hydrogen chloride produced then recombine further up the test tube forming ammonium chloride again.

This is shown in the equation below.

$$NH_4Cl(s) \rightleftharpoons NH_3(g) + HCl(g)$$

In fact most chemical reactions are reversible. In a reversible reaction the forward and reverse reactions occur at the same time and the reaction mixture contains both reactants and products.

Dynamic Equilibrium

If we consider the reversible reaction:

A + B $\rightleftharpoons$ C + D

At the beginning of the reaction, when reactants A and B are at their most concentrated, the rate of the forward reaction will be at its greatest. As the reaction proceeds A and B will be getting used up, so the concentrations of A and B are decreasing and so the rate of the forward reaction is also decreasing.

The products C and D are not present at the start and so, at the beginning, the rate of the reverse reaction will be zero. As the forward reaction proceeds the products C and D will be formed and their concentrations gradually increase. Therefore, the rate of the reverse reaction will increase. As the rate of the forward reaction decreases and the rate of the reverse reaction increases a balance point is eventually reached when the reaction **appears** to have stopped. In fact both forward and reverse reactions are still taking place but at the same speed. We say that the reactions at this point have reached **equilibrium**. Reversible reactions attain a state of **dynamic equilibrium** when the rates of forward and reverse reactions are equal.

The sign $\rightleftharpoons$ is used to show that a reaction is at dynamic equilibrium.

THE CONCEPT OF DYNAMIC EQUILIBRIUM contd

The changes in the rates of the forward and reverse reactions as equilibrium is being reached are summarised below.

A + B ⇌ C + D   Forward reaction much faster than reverse reaction

A + B ⇌ C + D   Forward reaction slowing down, reverse reaction getting faster

A + B ⇌ C + D   Reverse reaction almost as fast as forward reaction

A + B ⇌ C + D   Both reactions occurring at the same rate, dynamic equilibrium
has been reached

If the equilibrium concentrations of A and B are less than those of C and D, then the equilibrium position is said to lie to the right, that is to the side of the products.

If the equilibrium concentrations of A and B are greater than those of C and D, then the equilibrium position lies to the left, that is to the side of the reactants.

Position of Equilibrium

Although the rates of forward and reverse reactions are the same when equilibrium is reached, this does not mean that the concentrations of the reactants and products are the same. Usually at equilibrium there are more products than reactants or vice versa. Once a reaction system is at equilibrium, it is impossible to tell whether the equilibrium mixture was obtained by starting with the reactants or the products since the position of equilibrium is the same no matter from which direction it is approached.

DON'T FORGET

At equilibrium the concentrations of the reactants and the products remain constant although not necessarily equal.

LET'S THINK ABOUT THIS

Iodine dissolves in potassium iodide solution producing a brown solution.

Iodine dissolves in an organic solvent producing a purple solution.

If these two iodine solutions are added together then since the two solvents do not mix, an equilibrium is set up between the iodine in each of the solvents.

Test tube A contains iodine dissolved in potassium iodide solution (brown) to which colourless organic solvent is added. Iodine will start to move from the upper layer to the lower layer.

Test tube B contains iodine dissolved in an organic solvent (purple) to which colourless potassium iodide solution is added. Iodine will start to move from the lower layer to the upper layer.

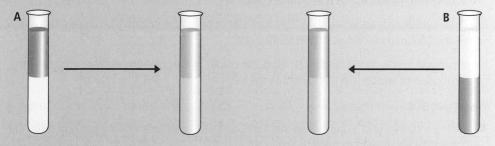

After some time the contents of both test tubes look identical. Both have a brown solution sitting on top of a purple solution.

Why has this happened?

A position of equilibrium has been reached. The rate at which the iodine is moving from the potassium iodide solution downwards to the organic solvent is the same rate at which it is moving upwards from the organic solvent to the potassium iodide solution. This experiment illustrates that the position of equilibrium is the same no matter from which direction it is approached.

EQUILIBRIUM 2

SHIFTING THE EQUILIBRIUM POSITION – THE EFFECT OF CHANGING CONCENTRATION AND PRESSURE

Changing the concentration

Consider the system at equilibrium which is present in a bottle of vinegar (dilute ethanoic acid):

$$CH_3COOH(aq) \rightleftharpoons CH_3COO^-(aq) + H^+(aq)$$

ethanoic acid molecules   ethanoate ions   hydrogen ions

Since the system is at equilibrium, the rate of the forward reaction is equal to the rate of the reverse reaction and the concentrations of the reactants and products are constant.

What happens if the concentration of ethanoate ions is changed? This can be done by adding solid sodium ethanoate to the equilibrium mixture.

The changes can be followed using universal indicator solution added to the mixture.

The universal indicator is an orange-red colour in ethanoic acid. When solid sodium ethanoate is added the universal indicator changes to a yellow-orange colour. This indicates that the solution has become less acidic, i.e. there are now fewer H^+ ions in the solution.

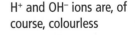

Looking at the equilibrium equation above, this means that if there are now fewer H^+ ions then the position of equilibrium must have shifted to the left so there are now more CH_3COOH molecules and fewer H^+ ions.

Now consider the equilibrium shown in the equation below:

$$Cr_2O_7^{2-}(aq) + OH^-(aq) \rightleftharpoons 2CrO_4^{2-}(aq) + H^+(aq)$$

orange dichromate ions   yellow chromate ions H^+ and OH^- ions are, of course, colourless

The equilibrium mixture is a yellow-orange colour.

If sodium hydroxide is added (adding OH^- ions), the colour of the solution becomes more yellow, so the equilibrium position has shifted to the right.

If hydrochloric acid is added (adding H^+ ions), the colour of the solution becomes more orange, so the position of equilibrium has shifted to the left.

We can say that:

- adding a species which appears on the right hand side of the equilibrium position shifts the position of equilibrium to the left, and

- adding a species which appears on the left hand side of the equilibrium position shifts the position of equilibrium to the right.

It is also true that:

- removing a species which appears on the right hand side of the equilibrium position shifts the position of equilibrium to the right, and

- removing a species which appears on the left hand side of the equilibrium position shifts the position of equilibrium to the left.

DON'T FORGET

To shift the position of equilibrium to the right (in order to make more of the product) we can:
- add more of one of the substances at the left hand side of the equation (the reactants) or
- remove the products as they are formed.

contd

SHIFTING THE EQUILIBRIUM POSITION – THE EFFECT OF CHANGING CONCENTRATION AND PRESSURE contd

Changing the pressure

Only if there are gases present, will changes in pressure affect the position of equilibrium.

Increasing the pressure shifts the position of equilibrium to the side of the equation which has the lower gas volume (lower number of moles of gas).

Decreasing the pressure shifts the position of equilibrium to the side of the equation which has the higher gas volume (higher number of moles of gas).

Consider the equilibrium equation given below:

$$N_2O_4(g) \rightleftharpoons 2NO_2(g)$$

light yellow	dark brown
1 mole of gas	2 moles of gas

Increasing the pressure shifts the position of equilibrium to the left since the volume of gas on the left side is less than the volume of gas on the right. Therefore, the colour of the gas mixture would become lighter when the pressure is increased.

Decreasing the pressure shifts the position of equilibrium to the right since the volume of gas on the right side is more than the volume of gas on the left. Therefore, the colour of the gas mixture would become darker when the pressure is decreased.

Another example is the preparation of methanol from synthesis gas as shown in the equilibrium equation:

$$CO(g) + 2H_2(g) \rightleftharpoons CH_3OH(g)$$

This equation shows three moles of gas on the left and only one mole of gas on the right.

Increasing the pressure shifts the position of equilibrium to the side of lower gas volume and therefore to the right. The preparation of methanol from synthesis gas gives a greater yield at high pressure.

> **DON'T FORGET**
>
> Changing the pressure only shifts the equilibrium position if there are **different numbers of moles of gas** on each side of the equilibrium equation.

LET'S THINK ABOUT THIS

1   A reaction involving a gas or gases will only come to equilibrium if it is carried out in a closed container. If the container is open then the gas(es) will escape and the system will not come to equilibrium.

Consider the equilibrium:

$$CaCO_3(s) \rightleftharpoons CaO(s) + CO_2(g)$$

If this is carried out in an open container, for example in a test tube, then the carbon dioxide gas will escape into the air and so the reverse reaction cannot take place. This means that the forward reaction will always be faster than the reverse reaction and so equilibrium will not be reached and eventually all the calcium carbonate will have decomposed into calcium oxide. Equilibrium will only be reached if the reaction is carried out in a closed container in which no carbon dioxide is allowed to escape. The reverse reaction can then take place and, as usual, equilibrium will be reached when the rate of the reverse reaction is the same as the rate of the forward reaction.

2   If the reaction is carried out in a closed container and equilibrium has been reached, what effect will increasing the pressure have on the position of equilibrium?

The equation $CaCO_3(s) \rightleftharpoons CaO(s) + CO_2(g)$ shows zero moles of gas on the left hand side and one mole of gas on the right hand side. Increasing the pressure shifts the position of equilibrium to the side with the lower gas volume and since zero is less than one, then the position of equilibrium will shift to the left hand side and there will be fewer molecules of carbon dioxide at this new equilibrium position.

EQUILIBRIUM 3

MORE ABOUT SHIFTING THE EQUILIBRIUM POSITION

Changing the temperature

For a chemical system in equilibrium an increase in temperature favours the endothermic reaction and a decrease in temperature favours the exothermic reaction.

Consider the equilibrium reaction, in which ΔH has a positive value. This means that the forward reaction is endothermic and so the reverse reaction must be exothermic. This is shown above and below the equilibrium arrow.

$$N_2O_4(g) \underset{\text{exothermic}}{\overset{\text{endothermic}}{\rightleftharpoons}} 2NO_2(g) \qquad \Delta H \text{ is positive}$$
light yellow   dark brown

If a sealed test tube containing this equilibrium mixture is placed in a beaker of hot water, the endothermic reaction is favoured and the position of equilibrium shifts to the right and the contents of the test tube become darker brown.

If the same sealed test tube containing this equilibrium mixture is placed in a beaker of iced water, the exothermic reaction is favoured and the position of equilibrium shifts to the left and the contents of the test tube become lighter in colour.

The effect of a catalyst on the position of equilibrium

Consider the potential energy diagram below for the following reversible reaction at equilibrium:

$$A + B \rightleftharpoons C + D$$

If a catalyst is added the rates of **both** the forward and reverse reactions will be increased.

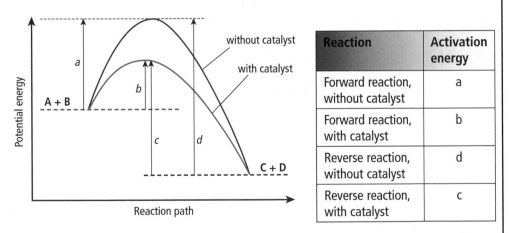

Reaction	Activation energy
Forward reaction, without catalyst	a
Forward reaction, with catalyst	b
Reverse reaction, without catalyst	d
Reverse reaction, with catalyst	c

The presence of the catalyst has lowered the activation energy for both the forward and reverse reactions by the same amount. Therefore, both forward and reverse reactions have been speeded up equally and so the position of equilibrium has not been altered.

Equilibrium in industry – the Haber Process

The manufacture of ammonia by the Haber Process is a very important industrial reaction. The reactants are nitrogen and hydrogen gases. The reaction is reversible and if the conditions were kept constant the following equilibrium would be attained.

$$N_2(g) + 3H_2(g) \rightleftharpoons 2NH_3(g) \qquad \Delta H = -92\,kJ\,mol^{-1}$$

Conditions are carefully chosen to provide a compromise between fast production and a high yield of ammonia and low costs. Factors which increase the rate of forward reactions are favourable.

DON'T FORGET

In a system at equilibrium an increase in temperature always favours the endothermic reaction while a decrease in temperature always favours the exothermic reaction.

DON'T FORGET

Catalysts have no effect on the position of equilibrium but they do speed up the rate of both forward and reverse reactions and so speed up the rate at which equilibrium is attained.

DON'T FORGET

The three factors which can change the position of equilibrium in a chemical reaction are concentration, temperature and pressure. Catalysts have no effect on the position of equilibrium.

contd

MORE ABOUT SHIFTING THE EQUILIBRIUM POSITION contd

Pressure

Since both the reactants and products are gases and the number of moles of gaseous products is less than the number of moles of gaseous reactants, a high pressure will shift the position of equilibrium to the side of the products. The pressure chosen is between 50 and 250 atmospheres. Higher pressures could be used but the relative increase in yield would not justify the increased costs.

Temperature

Since the forward reaction is exothermic (ΔH is negative), lowering the temperature will favour the forward reaction. However, the reaction is too slow at low temperatures and a compromise between a slow reaction at low temperatures and low yield at high temperatures has to be reached. As a result, the Haber Process is carried out at a moderate temperature of approximately 400–500°C. If the temperature is much higher than this the faster reaction will not compensate for the much lower yield.

Recycling of unused gases

The yield in the Haber Process, in fact, is only about 15%. The mixture of ammonia produced and the unreacted nitrogen and hydrogen is then cooled down and the ammonia gas condenses into ammonia liquid which is then tapped off. The unreacted nitrogen and hydrogen are then recirculated back through the reaction process.

Removal of product

Removal of the ammonia formed ensures that the rate of the reverse reaction is never equal to the rate of the forward reaction and so the system is not allowed to reach equilibrium. It is not in the manufacturer's interests for the ammonia formed to change back into nitrogen and hydrogen.

Catalyst

A catalyst lowers the activation energy for both the forward and reverse reactions. Using a catalyst therefore increases the rates of both the forward and reverse reactions. This allows the process to be carried out more quickly at a lower temperature. An iron catalyst is used in the Haber Process.

In industry the best practice is usually a compromise between factors such as yield, rate of reaction and operating costs. A catalyst will speed up the reaction at a lower temperature and so reduce costs but will have no effect on the yield. However the product will be made more quickly.

LET'S THINK ABOUT THIS

Another important industrial process is the Contact Process for the manufacture of sulphuric acid. The equation for one stage in the reaction is:

$$2SO_2(g) + O_2(g) \rightleftharpoons 2SO_3(g)$$

In this equilibrium reaction again the reactants and product are all gases. The total number of moles of gaseous reactants is greater than the number of moles of gaseous product. Therefore, an increase in pressure would shift the position of equilibrium to the right.

However, the Contact Process is carried out at a pressure only slightly above atmospheric pressure rather than the much higher pressure used in the Haber Process.

The reason that high pressure is used in the Haber Process is to increase the very low yield. The yield in the Contact Process is already over 90% at atmospheric pressure and so the extra costs of the high pressure would not be justified since the yield would only increase marginally.

ACIDS AND BASES 1

THE pH SCALE

What is the pH scale?

The pH scale is a continuous range from below 0 to above 14. Most solutions that you use in the chemical laboratory have pH values between 0 and 14 but certain solutions like fairly concentrated acids may even have negative pH values and some concentrated alkalis may have pH values above 14. The fact that it is a continuous scale means that the pH value can be non-integral, that is 3·1 or 12·6 and so on.

You will remember from earlier years that acids have pH values below 7, alkalis have pH values above 7 and that neutral solutions have pH = 7.

The pH of an aqueous solution is a measure of the concentration of hydrogen ions in that solution.

Consider an aqueous solution, which has a hydrogen ion concentration of 0·1 mol l^{-1}, for example 0·1 mol l^{-1} HCl.

$[H^+(aq)] = 0·1 = 1 \times 10^{-1}$ mol l^{-1}. This solution has a pH = 1.

[ ] means 'the concentration of' so [H$^+$] means the concentration of H$^+$ ions.

When this solution is diluted by a factor of 10, then its hydrogen ion concentration becomes 0·01 mol l^{-1} or $[H^+(aq)] = 0·01 = 1 \times 10^{-2}$ mol l^{-1}. This diluted solution has pH = 2.

These calculations show that the relationship between hydrogen ion concentration and pH is:

pH = −(power to which the hydrogen ion concentration is raised)

Equilibrium in water

Pure water is a very poor conductor of electricity. But the fact that it does conduct, even to a slight extent, shows that ions must be present.

In fact, in water there is an equilibrium between water molecules and hydrogen and hydroxide ions.

$$H_2O(l) \rightleftharpoons H^+(aq) + OH^-(aq)$$

The position of equilibrium lies well over to the left, i.e. there are many more water molecules present at equilibrium than there are hydrogen ions and hydroxide ions.

In water and other neutral aqueous solutions at 25°C the pH = 7.

This means that $[H^+(aq)] = 1 \times 10^{-7}$ mol l^{-1}

In water and neutral solutions, $[H^+(aq)] = [OH^-(aq)]$ and so $[OH^-(aq)]$ is also 1×10^{-7} mol l^{-1}

Therefore at 25°C, $\mathbf{[H^+(aq)] \times [OH^-(aq)] = 1 \times 10^{-14}}$ (from $1 \times 10^{-7} \times 1 \times 10^{-7}$)

This equation is true for water and for all aqueous solutions such as dilute acids and alkalis at 25°C.

DON'T FORGET

The lower the pH then the more acidic the solution.

contd

THE pH SCALE contd

Calculating the pH

We can use the relationship $[H^+(aq)] \times [OH^-(aq)] =$ 1×10^{-14} to work out the concentration of hydroxide ions in a solution given the concentration of hydrogen ions and vice-versa. This is shown in the table opposite. If we know the hydrogen ion concentration, $[H^+]$, then the pH can be calculated.

Estimating the pH from the hydrogen ion concentration

If you are given the hydrogen ion concentration in the form of $1 \times 10^{-x}\,mol\,l^{-1}$ then the pH = x where x is a whole number. However the pH value is not always a whole number, that is it can be non-integral. In this type of situation, sometimes we are asked to estimate the pH value, rather than calculate a definite value.

An example would be where we are asked to estimate the pH of $0.04\,mol\,l^{-1}$ NaOH.

First we would calculate $[H^+]$ from $[H^+(aq)] \times [OH^-(aq)] = 1 \times 10^{-14}$.

$[OH^-] = 0.04\,mol\,l^{-1}$ and so $[H^+] = \dfrac{1 \times 10^{-14}}{0.04} = 2.5 \times 10^{-13}\,mol\,l^{-1}$.

Since 2.5 is between 1 and 10, then the $[H^+]$ is between 1×10^{-13} and 10×10^{-13} (1×10^{-12}) $mol\,l^{-1}$ and so the pH must lie between 13 and 12.

Estimating the hydrogen ion concentration from the pH

If you are given a non-integral pH you may be asked to estimate the hydrogen ion concentration. An example would be where you are asked to estimate the hydrogen ion concentration in a solution which has pH = 3.8.

pH 3.8 is between pH = 3 and pH = 4 and so the hydrogen ion concentration, $[H^+]$ must be between 1×10^{-3} and $1 \times 10^{-4}\,mol\,l^{-1}$.

pH	$[H^+(aq)]/mol\,l^{-1}$		$[OH^-(aq)]/mol\,l^{-1}$	pH
14	1×10^{-14}		1	14
13	1×10^{-13}		1×10^{-1}	13
12	1×10^{-12}		1×10^{-2}	12
11	1×10^{-11}		1×10^{-3}	11
10	1×10^{-10}	Increasing alkalinity	1×10^{-4}	10
9	1×10^{-9}		1×10^{-5}	9
8	1×10^{-8}		1×10^{-6}	8
7	1×10^{-7}	Neutral	1×10^{-7}	7
6	1×10^{-6}		1×10^{-8}	6
5	1×10^{-5}	Increasing acidity	1×10^{-9}	5
4	1×10^{-4}		1×10^{-10}	4
3	1×10^{-3}		1×10^{-11}	3
2	1×10^{-2}		1×10^{-12}	2
1	1×10^{-1}		1×10^{-13}	1
0	1		1×10^{-14}	0

DON'T FORGET

If you know $[H^+]$ then you can work out $[OH^-]$ and vice-versa.

LET'S THINK ABOUT THIS

1 A trout fishery owner added limestone to his loch to combat the effects of acid rain. He managed to raise the pH of the water from 4 to 6.

The concentration of the H^+ (aq)

A increased by a factor of 2

B increased by a factor of 100

C decreased by a factor of 2

D decreased by a factor of 100.

2 The pH of a solution of hydrochloric acid was found to be 2.5.

The concentration of the H^+ (aq) ions in the acid must be

A greater than $0.1\,mol\,l^{-1}$

B between 0.1 and $0.01\,mol\,l^{-1}$

C between 0.01 and $0.001\,mol\,l^{-1}$

D less than $0.001\,mol\,l^{-1}$.

3 A lemon juice is found to have a pH of 3 and an apple juice a pH of 5.

From this information, the concentrations of H^+ (aq) ions in the lemon juice and apple juice are in the proportion (ratio)

A 100 : 1

B 1 : 100

C 20 : 1

D 3 : 5.

For answers, see p108.

ACIDS AND BASES 2

THE CONCEPT OF STRONG AND WEAK

Strong and weak acids

The words **strong** and **weak** have a quite different meaning from concentrated and dilute. A strong acid can be either concentrated or dilute depending on the number of moles of acid per litre of solution. The same applies to a weak acid.

A strong acid is completely dissociated (ionised) in aqueous solution.

For example $HCl(aq) \longrightarrow H^+(aq) + Cl^-(aq)$

A solution of hydrogen chloride, ie hydrochloric acid, will exist entirely as hydrogen ions and chloride ions – there will be no hydrogen chloride molecules present. The position of equilibrium lies completely over to the products side. Other strong acids include sulphuric and nitric acids.

A weak acid is only partially dissociated (ionised) in aqueous solution.

For example, ethanoic acid:

$$CH_3COOH(aq) \rightleftharpoons CH_3COO^-(aq) + H^+(aq)$$

The position of equilibrium lies well over to the left hand side and at equilibrium there will be few ethanoate and hydrogen ions compared to the much larger number of ethanoic acid molecules. There will be a greater concentration of hydrogen ions compared to hydroxide ions so ethanoic acid solution is acidic but weak compared to hydrochloric acid solution.

All carboxylic acids are weak acids, for example methanoic acid and propanoic acid. Other weak acids include carbonic acid, H_2CO_3 (carbon dioxide dissolved in water) and sulphurous acid, H_2SO_3 (sulphur dioxide dissolved in water).

$$H_2CO_3(aq) \rightleftharpoons 2H^+(aq) + CO_3^{2-}(aq)$$

$$H_2SO_3(aq) \rightleftharpoons 2H^+(aq) + SO_3^{2-}(aq)$$

DON'T FORGET

In aqueous solution, strong acids are completely ionised, but weak acids are only partially ionised.

Comparing strong and weak acids

Solutions of strong and weak acids which have the same concentration differ in pH, conductivity, and reaction rates. Results of experiments showing this are in the table below.

	0·1 mol l⁻¹ HCl (a strong acid)	0·1 mol l⁻¹ CH₃COOH (a weak acid)
pH	1·0	2·9
Conductivity (siemens)	$4·1 \times 10^{-2}$	$2·8 \times 10^{-4}$
Rate of reaction with Mg	fast	slow

The greater the concentration of H^+ ions then the lower is the pH. Conductivity is related to the number of ions – the more ions present then the higher is the conductivity. Magnesium reacts with the hydrogen ions in an acid and the greater the concentration of hydrogen ions then the faster is the rate of the reaction.

The results in the table show that there are many more hydrogen ions present in a strong acid compared to a weak acid when comparing acids of the same concentration. (Acids which have the same concentration are said to be equimolar).

However, if you were to carry out a titration neutralising, say, 25 cm³ of 0·1 mol l⁻¹ NaOH(aq) you would find that equal volumes of equimolar solutions of both hydrochloric acid and ethanoic acid would be required. For example, you would need 25 cm³ of 0·1 mol l⁻¹ HCl or 25 cm³ of 0·1 mol l⁻¹ CH₃COOH. We say that strong and weak acids do not differ in the stoichiometry of reactions.

contd

THE CONCEPT OF STRONG AND WEAK contd

Strong and weak bases

The same definitions apply to strong and weak bases as to strong and weak acids.

For example, a strong base is completely ionised in aqueous solution, that is all the available hydroxide ions are released into solution. Sodium hydroxide is a strong base.

$$Na^+OH^-(s) + H_2O(l) \longrightarrow Na^+(aq) + OH^-(aq)$$

A weak base is only partially ionised in aqueous solution. A weak base is made up of molecules. Only some of the molecules are dissociated (ionised) when dissolved in water. Ammonia is a weak base.

$$NH_3(aq) + H_2O(l) \rightleftharpoons NH_4^+(aq) + OH^-(aq)$$

The position of equilibrium lies well over to the left hand side and at equilibrium there will be few ammonium and hydroxide ions compared to the much larger number of ammonia molecules.

There will be a greater concentration of hydroxide ions compared to hydrogen ions so aqueous ammonia is alkaline but weak compared to sodium hydroxide solution.

Comparing strong and weak bases

Solutions of strong and weak bases which have the same concentration differ in pH and conductivity. Results of experiments showing this are in the table below.

	$0.1\,mol\,l^{-1}$ NaOH(aq) (a strong base)	$0.1\,mol\,l^{-1}$ NH_3(aq) (a weak base)
pH	13	11
Conductivity (siemens)	1.8×10^{-3}	6.2×10^{-5}

However, just as with strong and weak acids, strong and weak bases do not differ in the stoichiometry of their reactions. So, for example, $20\,cm^3$ of $0.1\,mol\,l^{-1}$ HCl(aq) would be neutralised by $10\,cm^3$ of $0.2\,mol\,l^{-1}$ NaOH (aq) or $10\,cm^3$ of $0.2\,mol\,l^{-1}$ NH_3(aq).

DON'T FORGET

You need to know which acids and bases are strong and which ones are weak.

LET'S THINK ABOUT THIS

When we say that strong and weak acids do not differ in stoichiometry of reactions, this means that one mole of a weak acid such as ethanoic acid would react in the same mole ratio as one mole of a strong acid such as hydrochloric acid when reacting with alkalis or metal oxides or metal carbonates or with metals like magnesium. The mole ratio would be the same but the reaction rate may be different. This is because as the hydrogen ions from the weak acid are being used up, the molecules of the weak acid dissociate further to replace these hydrogen ions, so that eventually the weak acid will produce the same number of hydrogen ions as the strong acid.

1   Which of the following is the same for equal volumes of $0.1\,mol\,l^{-1}$ solutions of sodium hydroxide and ammonia?

   A   The pH of solution

   B   The mass of solute present

   C   The conductivity of solution

   D   The number of moles of hydrochloric acid needed for neutralisation

2   Which of the following statements is **true** about an aqueous solution of ammonia?

   A   It has a pH less than 7.

   B   It is completely ionised.

   C   It contains more hydroxide ions than hydrogen ions.

   D   It reacts with acids producing ammonia gas.

For answers, see p108.

ACIDS AND BASES 3

THE pH OF SALT SOLUTIONS

Soluble salts of strong acids and strong bases

A soluble salt of a strong acid and a strong base dissolves in water to form a neutral solution. A good example is sodium chloride or common salt. The parent acid is hydrochloric acid, a strong acid. The parent base is sodium hydroxide, a strong base.

Since both the parent acid and parent base are strong, they are fully ionised in water and the equilibrium present in water is unaffected.

$$H_2O(l) \rightleftharpoons H^+(aq) + OH^-(aq).$$

Since there are equal numbers of H^+ and OH^- ions the pH of the salt solution is 7.

Soluble salts of strong acids and weak bases

A solution of a salt of a strong acid and a weak base has a pH lower than 7, i.e. it is acidic.

Examples of salts from strong acids and weak bases include ammonium chloride and ammonium nitrate.

Consider a solution of ammonium chloride, $NH_4^+Cl^-$. This is the salt of the strong acid, hydrochloric, and the weak base, ammonia.

The equilibrium present in water is:

$$H_2O(l) \rightleftharpoons H^+(aq) + OH^-(aq).$$

In solution ammonium chloride is acidic and so contains more H^+ than OH^-. This means that some of the OH^- ions present in the water equilibrium must have been removed.

The positive NH_4^+ ions present in ammonium chloride have reacted with the negative OH^- ions to form ammonia molecules. This happens because ammonia is a weak base and in the equilibrium equation below the position of equilibrium lies well over to the right so that there will be mainly aqueous ammonia molecules and fewer ions.

$$NH_4^+(aq) + OH^-(aq) \rightleftharpoons NH_3(aq) + H_2O(l)$$

There are now fewer OH^- compared to H^+ ions and this will continue to be true even when more water molecules dissociate. Therefore the salt solution is acidic.

For any salt of a strong acid and weak base, $[H^+] > [OH^-]$ and so the solution is acidic.

To explain why it is acidic just work out which ions in the salt have removed some of the OH^- ions from the water. Since hydroxide ions are negative it must be the positive ions from the salt which remove the OH^- ions.

Soluble salts of weak acids and strong bases

A solution of a salt of a weak acid and strong base has a pH above 7, i.e. it is alkaline.

Examples of salts from weak acids and strong bases include sodium ethanoate and potassium carbonate.

Consider a solution of sodium ethanoate, $Na^+CH_3COO^-$. This is the salt of the weak acid, ethanoic, and the strong base, sodium hydroxide. Again, the equilibrium present in water is:

$$H_2O(l) \rightleftharpoons H^+(aq) + OH^-(aq).$$

In solution sodium ethanoate is alkaline and so contains more OH^- than H^+. This means that some of the H^+ ions present in the water equilibrium must have been removed.

contd

THE pH OF SALT SOLUTIONS contd

The negative CH_3COO^- ions present in sodium ethanoate have reacted with the positive H^+ ions to form ethanoic acid molecules. This happens because ethanoic acid is a weak acid and in the equilibrium equation below, the position of equilibrium lies well over to the right so that there will be mainly aqueous ethanoic acid molecules and fewer ions.

$$CH_3COO^-(aq) + H^+(aq) \rightleftharpoons CH_3COOH(aq)$$

There are now fewer H^+ compared to OH^- ions and this will continue to be true even when more water molecules dissociate. Therefore the salt solution is alkaline.

For any salt of a weak acid and strong base, $[H^+] < [OH^-]$ and so the solution is alkaline.

To explain why it is alkaline just work out which ions in the salt have removed some of the H^+ ions from the water. Since hydrogen ions are positive it must be the negative ions from the salt which remove the H^+ ions.

pH of soap solution

Soaps are salts of fatty acids which are weak acids and strong bases such as sodium hydroxide or potassium hydroxide. Soaps therefore dissolve in water to form alkaline solutions.

A typical salt is sodium stearate. This is the salt of stearic acid and sodium hydroxide. The formula of stearic acid is $C_{17}H_{35}COOH$ and the formula of sodium stearate is $Na^+C_{17}H_{35}COO^-$.

DON'T FORGET

The pH of a salt solution depends on its parent acid and parent base. If both are strong then the salt solution is neutral. If the parent acid is strong and the parent base is weak, then the salt solution will be acidic. If the parent acid is weak and the parent base is strong, then the salt solution will be alkaline.

LET'S THINK ABOUT THIS

1   Propanoic acid can be used to prepare the salt potassium propanoate, CH_3CH_2COOK.

Explain why potassium propanoate solution has a pH greater than 7.

In your answer you should mention the **two** equilibria involved.

What are the two equilibria referred to in the question?

They are the equilibrium present in water and the equilibrium in propanoic acid solution.

Since the salt solution has pH greater than 7, then it is the salt of a weak acid and strong base. The strong base is NaOH and since it is fully ionised we cannot write an equilibrium equation for its dissociation.

Propanoic acid is the weak acid and the equilibrium equation for its dissociation is:

$$CH_3CH_2COOH(aq) \rightleftharpoons CH_3CH_2COO^-(aq) + H^+(aq).$$

The equation showing the equilibrium present in water is:

$$H_2O(l) \rightleftharpoons H^+(aq) + OH^-(aq).$$

In both equations the position of equilibrium lies well over to the left, ie to the reactants side. When sodium propanoate dissolves in water, the propanoate ions react with the hydrogen ions from the water forming propanoic acid molecules. This effectively removes some of the $H^+(aq)$ ions and so there will be less hydrogen ions than hydroxide ions present and the solution is alkaline.

2   Potassium sorbate is a salt that is used as a preservative in margarine. Potassium sorbate dissolves in water to form an alkaline solution.

What does this indicate about sorbic acid?

Since potassium sorbate forms an alkaline solution in water, then it is the salt of a weak acid and strong base. Therefore sorbic acid must be a weak acid. (It is not necessary even to know its formula or write an equation for its dissociation to be able to state that it must be a weak acid.)

REDOX REACTIONS 1

OXIDISING AND REDUCING AGENTS

Ion-electron equations

Ion-electron equations are used to describe the oxidation and reduction processes that make up a redox reaction. In an oxidation ion-electron equation, the electrons appear on the products side because oxidation involves a loss of electrons. In a reduction ion-electron equation they are on the reactants side since reduction involves a gain of electrons. In any ion-electron equation, the net charge on the reactants side must always equal that on the products side.

To illustrate these ideas, consider the reaction between iron(II) sulphate solution and an acidified potassium dichromate solution in which iron(III) ions and chromium(III) ions are formed.

The iron(II) ions have been converted into iron(III) ions, i.e. $Fe^{2+}(aq) \longrightarrow Fe^{3+}(aq)$. To balance the charges, an electron has to be added to the product side giving a net charge of 2+ on each side:

$$Fe^{2+}(aq) \longrightarrow Fe^{3+}(aq) + e^-$$

Since the electrons appear on the products side, the iron(II) ions have been oxidised. This implies that the dichromate ions in the acidified potassium dichromate solution must have been reduced to chromium(III) ions, that is $Cr_2O_7^{2-}(aq) \longrightarrow Cr^{3+}(aq)$.

Working out the ion-electron equation for this particular reduction process is not quite as straight forward as that for the oxidation but it can be achieved by applying the following rules:

1   Balance the non-oxygen element, in this case the chromium:

$$Cr_2O_7^{2-}(aq) \longrightarrow 2Cr^{3+}(aq)$$

2   Balance the oxygen. Oxygen is needed on the product side and is introduced as water molecules. Since the dichromate ions contain seven O atoms then seven H_2O molecules will be required:

$$Cr_2O_7^{2-}(aq) \longrightarrow 2Cr^{3+}(aq) + 7H_2O(l)$$

3   Balance the hydrogen. Hydrogen is needed on the reactants side and is introduced as hydrogen ions. 14 H^+ ions are required:

$$Cr_2O_7^{2-}(aq) + 14H^+(aq) \longrightarrow 2Cr^{3+}(aq) + 7H_2O(l)$$

4   Finally balance the charge. As it stands, the net charge on the reactants side is 12+ (14+ from the 14 H^+ ions and 2− from the one $Cr_2O_7^{2-}$ ion). On the products side, the net charge is 6+ from the two Cr^{3+} ions. To balance the charge we introduce electrons. Six of them are needed on the reactants side to make the net charge on each side 6+:

$$Cr_2O_7^{2-}(aq) + 14H^+(aq) + 6e^- \longrightarrow 2Cr^{3+}(aq) + 7H_2O(l)$$

The electrons appear on the reactants side confirming that it is a reduction process.

You will notice that hydrogen ions are needed for dichromate ions to be reduced and this explains why the potassium dichromate solution initially present had to be acidified.

Many ion-electron equations you will need can be found on page 11 of your Data Booklet. They are all written as reductions and so to write an oxidation ion-electron equation all you have to do is simply reverse the relevant reduction equation. There will be occasions when you are asked to write an ion-electron equation which doesn't appear on page 11. In these cases, you must derive it from first principles using the rules outlined above.

DON'T FORGET

Given reactant and product species, you must be able to write ion-electron equations which include H^+ ions and H_2O molecules.

contd

OXIDISING AND REDUCING AGENTS contd

Redox equations

Balanced equations for redox reactions can be written by combining the ion-electron equations for the oxidation and reduction processes. Consider the redox reaction between sodium sulphite solution and acidified potassium permanganate solution. The relevant ion-electron equations are:

oxidation: $SO_3^{2-}(aq) + H_2O(l) \longrightarrow SO_4^{2-}(aq) + 2H^+(aq) + 2e^-$

reduction: $MnO_4^-(aq) + 8H^+(aq) + 5e^- \longrightarrow Mn^{2+}(aq) + 4H_2O(l)$

The redox equation has to be balanced with respect to the electrons, that is the number of electrons lost in the oxidation process must equal the number gained in the reduction process. This implies that we must multiply the oxidation equation by five and the reduction equation by two:

$$5SO_3^{2-}(aq) + 5H_2O(l) \longrightarrow 5SO_4^{2-}(aq) + 10H^+(aq) + 10e^-$$
$$2MnO_4^-(aq) + 16H^+(aq) + 10e^- \longrightarrow 2Mn^{2+}(aq) + 8H_2O(l)$$

With the electrons balanced, we add these ion-electron equations together giving:

$$5SO_3^{2-}(aq) + 5H_2O(l) + 2MnO_4^-(aq) + 16H^+(aq) \longrightarrow 5SO_4^{2-}(aq) + 10H^+(aq) + 2Mn^{2+}(aq) + 8H_2O(l)$$

No electrons appear in the equation because they cancel out. You will notice, however, that H^+ ions and H_2O molecules appear on both sides of the equation and these too have to be cancelled down. We finally arrive at the balanced redox equation:

$$5SO_3^{2-}(aq) + 2MnO_4^-(aq) + 6H^+(aq) \longrightarrow 5SO_4^{2-}(aq) + 2Mn^{2+}(aq) + 3H_2O(l)$$

It is important to remember that neither electrons nor spectator ions nor spectator molecules must feature in redox equations.

Oxidising and reducing agents

When a substance is oxidised it loses or donates electrons and for this to happen some other substance has to gain or accept these electrons and be reduced. The **electron-acceptor** is known as the **oxidising agent** and the **electron-donor** is called the **reducing agent**. Let's take a typical redox reaction:

$$2Mg(s) + O_2(g) \longrightarrow 2MgO(s)$$

and see if we can identify the oxidising and reducing agents. To do this we need to split up the redox equation into its oxidation and reduction ion-electron equations. This is straight forward provided you recognise that magnesium oxide is ionic and is made up of Mg^{2+} and O^{2-} ions. The Mg atoms have been converted into Mg^{2+} ions and so the relevant ion electron-equation must be:

$$Mg(s) \longrightarrow Mg^{2+}(s) + 2e^-$$

We can see that the Mg atoms have been oxidised and so the O_2 molecules must have been reduced:

$$O_2(g) + 4e^- \longrightarrow 2O^{2-}(s)$$

Since the oxygen molecules are accepting electrons then they must be the oxidising agents. This implies that the magnesium atoms must be the reducing agents and this is confirmed by the fact that they are donating electrons.

> **DON'T FORGET**
>
> An oxidising agent is an electron-acceptor and a reducing agent is an electron-donor.

LET'S THINK ABOUT THIS

Chlorofluorocarbons (CFCs) were banned because they destroy the ozone layer (see page 58) and a safe way of disposing stockpiled CFCs had to be found. A redox reaction came to the rescue. In the process, the CFCs are vaporised and passed through a layer of powdered sodium oxalate at 270°C. Using CF_2Cl_2 as a typical CFC, the equation for the reaction is:

$$CF_2Cl_2(g) + 2(Na^+)_2C_2O_4^{2-}(s) \longrightarrow 2Na^+F^-(s) + 2Na^+Cl^-(s) + C(s) + 4CO_2(g).$$

If this **is** a redox reaction, then we should be able to write the corresponding oxidation and reduction ion-electron equations. You will notice that the Na^+ ions are spectator ions and that the oxalate ions ($C_2O_4^{2-}$) have been converted into carbon dioxide, $2C_2O_4^{2-} \longrightarrow 4CO_2(g)$. In order to balance the charge four electrons need to be added to the product side giving:

$$2C_2O_4^{2-} \longrightarrow 4CO_2(g) + 4e^-.$$

The oxalate ions have therefore been oxidised and so the CF_2Cl_2 molecules must have been reduced:

$$CF_2Cl_2(g) + 4e^- \longrightarrow 2F^-(s) + 2Cl^-(s) + C(s).$$

REDOX REACTIONS 2

REDOX TITRATIONS

Titration

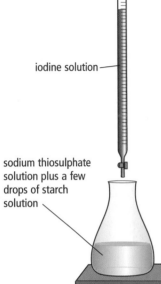

iodine solution

sodium thiosulphate solution plus a few drops of starch solution

Titration is an experimental procedure which can be used to determine the accurate concentration of a solution. It involves adding a solution of known concentration, ie a standard solution, to one of unknown concentration until reaction between the two is just complete. An indicator is usually added to the reaction mixture to show the end-point of the titration.

You are already familiar with acid/base titrations but the same procedure can equally well be used with oxidising and reducing agents. Titrating an oxidising agent against a reducing agent or vice versa is called a **redox titration**. To illustrate the redox titration technique, let's consider a particular example. Suppose we wanted to determine the concentration of a sodium thiosulphate $(Na_2S_2O_3)$ solution. Sodium thiosulphate is a reducing agent and can be determined by titrating it against a standard solution of an oxidising agent. Iodine solution is a suitable oxidising agent in this case and the redox equation for the reaction is:

$$2S_2O_3^{2-}(aq) + I_2(aq) \longrightarrow S_4O_6^{2-}(aq) + 2I^-(aq).$$

$25 \cdot 0\,cm^3$ of the sodium thiosulphate solution is pipetted into a conical flask and a few drops of starch indicator solution are added. The brown iodine solution is carefully added from a burette to the colourless sodium thiosulphate solution in the flask. The thiosulphate ions reduce the iodine molecules to colourless iodide ions and before the end-point is reached, the solution in the flask remains colourless. The titration is continued to the end-point which is marked by the sudden appearance of a blue-black colour. The colour change occurs because, at this point, no thiosulphate ions are left to reduce the iodine molecules and this leaves them free to react with the starch indicator to form the familiar blue-black coloured complex. The titrations are repeated until concordant results, ie two consecutive titre volumes within $0\cdot1\,cm^3$ of each other, are obtained. From the results, the concentration of the sodium thiosulphate solution can be calculated.

> ### DON'T FORGET
>
> In a redox titration, an oxidising agent is titrated against a reducing agent or vice versa.

Redox titration calculations

Example 1

A typical set of results for the above redox titration is outlined below:

Concentration of iodine solution = $0\cdot101\,mol\,l^{-1}$
Volume of sodium thiosulphate solution = $25\cdot0\,cm^3$

Titration	1	2	3
Initial burette reading (cm³)	1·1	19·7	0·9
Final burette reading (cm³)	19·7	37·8	19·1
Titre volume (cm³)	18·6	18·1	18·2

Average volume of iodine solution = $18\cdot15\,cm^3$ $(0\cdot01815\,litres)$

Notice that in calculating the average volume of iodine solution, the first titre volume is ignored since it is not concordant.

Knowing the concentration and volume of the iodine solution, the number of moles of iodine used in the titration can be worked out:

number of moles of iodine = $V \times c = 0\cdot01815 \times 0\cdot101 = 0\cdot001833\,mol$

contd

REDOX TITRATIONS contd

Using the balanced redox equation, the number of moles of sodium thiosulphate can now be determined:

$$2S_2O_3^{2-}(aq) \quad + \quad I_2 \longrightarrow S_4O_6^{2-}(aq) + 2I^-(aq)$$

$$2\,mol \longleftrightarrow 1\,mol$$

$$2 \times 0.001833 \longleftrightarrow 0.001833\,mol$$

$$= 0.003666\,mol$$

Now that we have the number of moles of sodium thiosulphate (0.003666) and its volume (25.0 cm^3 which is 0.0250 litres), we can finally calculate its concentration:

$$\text{Concentration of sodium thiosulphate solution} = \frac{n}{V} = \frac{0.003666}{0.0250} = 0.147\,mol\,l^{-1}$$

Example 2

Oxalic acid is found in rhubarb. The mass of oxalic acid in a carton of rhubarb juice can be found by titrating samples of the juice with a solution of potassium permanganate. The equation for the overall reaction is:

$$5(COOH)_2(aq) + 6H^+(aq) + 2MnO_4^-(aq) \longrightarrow 2Mn^{2+}(aq) + 10CO_2(g) + 8H_2O(l)$$

In an investigation using a 500 cm^3 carton of rhubarb juice, 25.0 cm^3 samples were titrated with 0.0400 mol l^{-1} potassium permanganate solution and the average titre volume was found to be 26.9 cm^3. Calculate the mass of oxalic acid in the 500 cm^3 carton of rhubarb juice assuming that one mole of oxalic acid has a mass of 90.0 g.

With such a lot of information, it is often difficult to know where to start such a calculation. A good idea is to begin with the titration results. We are given the volume and concentration of the potassium permanganate solution and from these we can work out the number of moles of potassium permanganate:

$$\text{number of moles of potassium permanganate} = V \times c = 0.0269 \times 0.0400 = 0.001076\,mol.$$

Using the balanced redox equation, we can then determine the number of moles of oxalic acid:

$$5(COOH)_2(aq) + 6H^+(aq) + 2MnO_4^-(aq) \longrightarrow 2Mn^{2+}(aq) + 10CO_2(g) + 8H_2O(l)$$

$$5\,mol \longleftrightarrow 2\,mol$$

$$5 \times \frac{0.001076}{2} \longleftrightarrow 0.001076\,mol$$

$$= 0.002690\,mol$$

So, 0.002690 mol of oxalic acid are present in 25 cm^3 of rhubarb juice which implies that in the 500 cm^3 carton there will be $0.002690 \times \frac{500}{25.0} = 0.0538$ mol of oxalic acid.

We are told that one mole of oxalic acid has a mass of 90.0 g, so the mass of oxalic acid in the 500 cm^3 carton = $0.0538 \times 90.0 = 4.84$ g.

LET'S THINK ABOUT THIS

Redox titrations are unusual in that in some cases there is no need to add a separate indicator since one of the reactants acts as its own indicator. Potassium permanganate is such an example. It serves as its own indicator because permanganate ions have a purple colour but when they are reduced, colourless manganese(II) ions are formed. So, in a titration involving potassium permanganate as in example 2 above, the end-point of the titration is marked by the colourless solution in the conical flask turning pink. It is pink rather than purple because the concentration of permanganate ions at the end-point is so low.

REDOX REACTIONS 3

ELECTROLYSIS

Qualitative electrolysis

Any molten liquid or aqueous solution that contains ions which are free to move and therefore carry a current of electricity is called an **electrolyte**. When a direct current is passed through an electrolyte, chemical changes occur at the electrodes which results in its decomposition. This process is known as **electrolysis**.

Consider the electrolysis of copper(II) chloride solution in the electrolytic cell shown opposite.

The positively charged copper(II) ions move to the negative electrode where they are reduced to copper atoms: $Cu^{2+}(aq) + 2e^- \longrightarrow Cu(s)$

The negatively charged chloride ions migrate to the positive electrode and are oxidised to chlorine molecules:

$$2Cl^-(aq) \longrightarrow Cl_2(g) + 2e^-$$

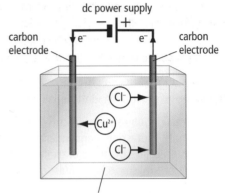

copper(II) chloride solution

The overall redox equation for the electrolysis is:

$$CuCl_2(aq) \longrightarrow Cu(s) + Cl_2(g)$$

It is important to note that:

- redox reactions associated with electrolysis do not occur of their own free will and require an input of electrical energy to bring them about,

- during electrolysis, oxidation always occurs at the positive electrode and reduction always occurs at the negative electrode.

Quantitative electrolysis

The quantity of electricity which passes through an electrolyte during electrolysis can be determined using the relationship:

$$Q = It$$

where
- **Q** is the quantity of electrical charge in coulombs (C)
- **I** is the current in amps (A)
- **t** is the period of time in seconds (s) during which the current passes.

How is this quantity of electrical charge related to the amount of product formed at each electrode during electrolysis? To help answer this question, consider the information in the table below which shows the number of coulombs required to produced one mole of each of the elements silver, copper, aluminium and oxygen and the relevant ion-electron equations involved in their formation:

Element	Quantity of electrical charge required to produce 1 mole (C)	Ion-electron equation
silver	96 500 or 1 × 96 500	$Ag^+ + 1e^- \longrightarrow Ag$
copper	193 000 or 2 × 96 500	$Cu^{2+} + 2e^- \longrightarrow Cu$
aluminium	289 500 or 3 × 96 500	$Al^{3+} + 3e^- \longrightarrow Cu$
oxygen	386 000 or 4 × 96 500	$2H_2O \longrightarrow O_2 + 4H^+ + 4e^-$

The relationship is now apparent – the quantity of electrical charge required to produce one mole of an element is **n × 96 500 C** where **n** is the number of moles of electrons in the relevant ion-electron equation.

contd

ELECTROLYSIS contd

The data in the above table also reveal a link between electrons and electrical charge. They show that the electrical charge associated with one mole of electrons is a constant and takes the value $96\,500\,C\,mol^{-1}$ or $9.65 \times 10^4\,C\,mol^{-1}$. This constant is called the Faraday constant after Michael Faraday who did much of the pioneering work on electrolysis, and can be found on page 19 of the Data Booklet.

DON'T FORGET

$96\,500\,C$ ($9.65 \times 10^4\,C$) is the electrical charge associated with one mole of electrons.

Electrolysis calculations

Example 1

A current of $10.0\,A$ is passed through molten magnesium chloride for 15.0 minutes. Calculate the mass of magnesium deposited at the negative electrode.

Knowing the current and the time, we can work out the quantity of electrical charge passed through the molten magnesium chloride:

$$Q = It = 10.0 \times 15.0 \times 60 = 9000\,C$$

It is important to remember to convert the time into seconds.

The next stage is to write the ion-electron equation for the reduction of the magnesium ions at the negative electrode and proceed with the calculation:

$$
\begin{array}{ccc}
Mg^{2+} & + \quad 2e^- & \longrightarrow \quad Mg \\
& 2\,mol & 1\,mol \\
& 2 \times 96\,500\,C & \longleftrightarrow \quad 24.3\,g \\
& 9000\,C & \longleftrightarrow \quad 24.3 \times \dfrac{9000}{2 \times 96\,500} \\
& & = 1.13\,g
\end{array}
$$

Example 2

In an experiment in which dilute sulphuric acid was electrolysed $98.5\,cm^3$ of oxygen was produced at the positive electrode:

$$2H_2O \longrightarrow O_2 + 4H^+ + 4e^-$$

If a current of $12.6\,A$ was used, calculate the time it took to produce $98.5\,cm^3$ of oxygen. (Assume the molar volume of oxygen to be 24.3 litres mol^{-1}.)

In this calculation, we start by using the ion-electron equation to calculate the quantity of charge, Q, required in producing $98.5\,cm^3$ (0.0985 litres) of oxygen:

$$
\begin{array}{ccccc}
2H_2O & \longrightarrow & O_2 & + \quad 4H^+ & + \quad 4e^- \\
& & 1\,mol & & 4\,mol \\
& & 24.3\,litres & \longleftrightarrow & 4 \times 96500\,C \\
& & 0.0985\,litres & \longleftrightarrow & 4 \times 96500 \times \dfrac{0.0985}{24.3} \\
& & & & = 1565\,C
\end{array}
$$

Now that we have Q ($1565\,C$), we can rearrange the relationship $Q = It$ to find the time in seconds:

$$t = \frac{Q}{I} = \frac{1565}{12.6} = 124\,s$$

⚙ LET'S THINK ABOUT THIS

Copper extracted from its ores doesn't conduct electricity particularly well and only after it has been purified can it be used in making electrical wiring. The purification of copper on a commercial scale is achieved by electrolysis. The electrolytic cell has copper electrodes dipping into a solution of copper(II) sulphate. The impure copper forms the positive electrode while a piece of pure copper is used as the negative electrode. The ion-electron equations taking place in the cell are:

positive electrode (impure copper): $\quad Cu(s) \longrightarrow Cu^{2+}(aq) + 2e^-$

negative electrode (pure copper): $\quad Cu^{2+}(aq) + 2e^- \longrightarrow Cu(s)$

Overall, the net reaction is $Cu(s) \longrightarrow Cu(s)$ making it appear that nothing is happening. However, something is taking place in the cell – the copper is being transferred from the impure positive electrode to the pure negative electrode.

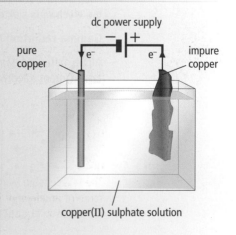

dc power supply

pure copper — impure copper

copper(II) sulphate solution

NUCLEAR CHEMISTRY 1

TYPES OF RADIATION

Radioactivity

Some isotopes of some elements are unstable and they spontaneously breakdown or decay into stable atoms with the emission of radiation and with the release of energy. This phenomenon is known as **radioactivity**.

In radioactive decay, changes take place in the nuclei of radioactive isotopes, or **radioisotopes**. In other words, radioactive decay involves nuclear reactions. This is unlike ordinary chemical reactions where the nuclei of the atoms remain intact and only the outer electrons are involved.

The stability of an atom depends on the relative numbers of neutrons and protons in its nucleus. When the number of neutrons is plotted against the number of protons for all stable naturally occurring nuclei, it is evident that there is a band of stability. We can see that the lighter stable nuclei have approximately equal numbers of neutrons and protons but as the nuclei get heavier, the number of neutrons increases more rapidly than the number of protons. Nuclei that lie outwith the stability band are radioactive and as they decay these unstable nuclei will emit radiation and their neutron to proton ratio will change. The decay process will continue until stable nuclei, that is nuclei with neutron to proton ratios which lie inside the stability band, are formed.

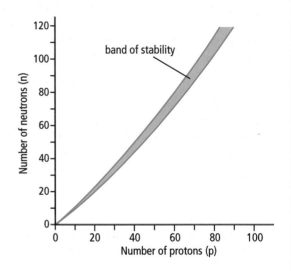

DON'T FORGET

The stability of nuclei depends on their neutron to proton ratio.

The nature and properties of radiation

There are three types of radiation – **alpha** (α), **beta** (β) and **gamma** (γ). Alpha and beta radiations are made up of particles but gamma radiation is made up of electromagnetic waves.

Alpha particles are **helium nuclei** and each has a charge of 2+ and contains two protons and two neutrons. The symbol for an alpha particle is ^4_2He.

Beta particles are high-energy **electrons** each with a charge of –1 and the symbol, $^0_{-1}\text{e}$. It seems strange that a nucleus could emit an electron since nuclei contain no electrons. What happens is that a neutron breaks up into a proton and an electron and as soon as it is formed, the electron is ejected at high speed from the nucleus as a beta particle: $^1_0\text{n} \longrightarrow {}^1_1\text{p} + {}^0_{-1}\text{e}$.

Gamma radiation has no mass or charge and has the symbol, γ.

Alpha, beta and gamma radiations have different penetrating powers as can be seen from this diagram. Clearly, gamma radiation is the most penetrating with alpha being the least penetrating.

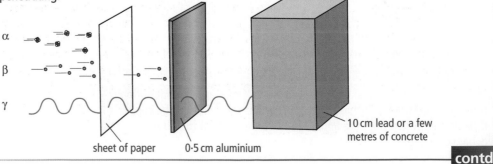

contd

TYPES OF RADIATION contd

Nuclear reactions

Nuclear reactions take place when radioisotopes decay. The changes that take place in their nuclei depend on the type of radiation that is emitted:

- **Alpha decay**
  When a nucleus emits an alpha particle, ie a particle containing two protons and two neutrons, its atomic number will decrease by two (loss of two protons) and its mass number will decrease by four (loss of two protons and two neutrons). Uranium-238, for example, decays by alpha emission and the nuclear equation for the decay is:

$$^{238}_{92}U \longrightarrow {}^{234}_{90}Th + {}^{4}_{2}He$$

- **Beta decay**
  A beta particle is formed along with a proton when a neutron in a radioactive nucleus breaks up. As a result of beta decay, the atomic number of the nucleus will increase by one since it has one more proton than it had originally. The mass number, however, will not change since a neutron has been replaced by a proton and they have the same mass. Bismuth-212, for example, decays by beta emission:

$$^{212}_{83}Bi \longrightarrow {}^{212}_{84}Po + {}^{0}_{-1}e$$

- **Gamma decay**
  Since gamma rays have no mass and no charge, their emission will have no effect on the atomic number or mass number of the radioisotope. Cobalt-60, for example, decays by gamma emission:

$$^{60}_{27}Co \longrightarrow {}^{60}_{27}Co + {}^{0}_{0}\gamma$$

Other nuclear reactions include the following:

- **Neutron capture**
  When nitrogen-14 nuclei are bombarded with neutrons, each 'captures' one of those neutrons to form carbon-14 and a proton:

$$^{14}_{7}N + {}^{1}_{0}n \longrightarrow {}^{14}_{6}C + {}^{1}_{1}p$$

- **Proton capture**
  Nitrogen-14 nuclei can also undergo proton capture when they are bombarded with protons. In this case, oxygen-15 is formed:

$$^{14}_{7}N + {}^{1}_{1}p \longrightarrow {}^{15}_{8}O$$

Notice that in balanced nuclear equations, like those above:

total mass number on reactants side = total mass number on products side
total atomic number on reactants side = total atomic number on products side.

> **DON'T FORGET**
>
> You must be able to write balanced nuclear equations involving neutrons, protons, alpha particles and beta particles.

LET'S THINK ABOUT THIS

The transmutation of elements, the conversion of one element into another, particularly base metals like lead into precious metals like gold, was the alchemists' dream. Having access only to ordinary chemical reactions, their attempts proved fruitless. However, with the discovery of radiation in 1896, natural transmutation was recognised but it wasn't until 1919 that artificial transmutation was achieved when Rutherford converted nitrogen into oxygen. He bombarded a sample of nitrogen with alpha particles and oxygen-17 isotopes were produced along with protons:

$$^{14}_{7}N + {}^{4}_{2}He \longrightarrow {}^{17}_{8}O + {}^{1}_{1}p$$

NUCLEAR CHEMISTRY 2

HALF-LIVES

What is half-life?

When a sample of a radioisotope decays, individual nuclei within that sample do not disintegrate at regular intervals. The disintegration is completely **random**. Nevertheless, when these random events are 'averaged out', the decay of a radioisotope follows a definite pattern as illustrated in the following graph:

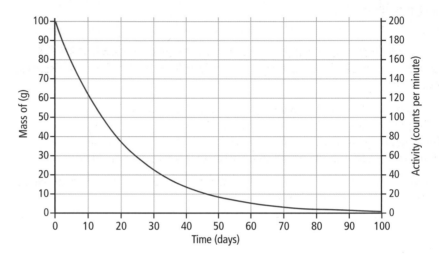

Although this graph shows the decay curve for phosphorus-32, decay curves for all other radioisotopes take a similar form.

We can see that the rate decreases with increasing time just as it does in ordinary chemical reactions. However, unlike ordinary chemical reactions, the rate of decay is not affected by changes in temperature.

From the decay curve we can work out that it takes approximately 14 days for the mass of phosphorus-32 to fall from 100 g to 50 g, i.e. to halve, and for the activity to fall from 200 counts per minute (cpm) to 100 cpm, ie to halve. This period of time is known as the half-life (often abbreviated to $t_{1/2}$) of phosphorus-32. Radioisotopes have different half-lives. For example, radon-220 has a half-life of only 55 s while uranium-238 has a half-life of 4.51×10^9 years.

It is important to note that while the activity of a radioisotope depends on the mass present, the half-life is independent of its mass. For example, the radiation given off by a 10 g sample of phosphorus-32 will be ten times that given off by a 1 g sample but the half-life of phosphorus-32 is 14 days no matter whether you have 10 g of it or just 1 g. It is also important to remember that the half-life of a radioisotope is the same no matter whether it is present as atoms or as ions. For example, $^{32}_{15}P$ atoms and $^{32}_{15}P^{3-}$ ions have the same half-life. The reason for this is that the nuclei of an atom and its ion are identical.

Half-life calculations

Example 1

Strontium-90 is a hazardous radioactive isotope that resulted from nuclear testing and it has a half-life of 27 years. After 54 years, a sample of strontium-90 had a mass of 0·20 g. What was the mass of the original sample of strontium-90?

$$54 \text{ years represents } \frac{54}{27} = 2 \text{ half-life periods.}$$

After 54 years ($2 \times t_{1/2}$), mass of strontium-90 = 0·20 g
after 27 years ($1 \times t_{1/2}$), mass of strontium-90 = 2 × 0·20 g
after 0 years ($0 \times t_{1/2}$), mass of strontium-90 = 2 × 2 × 0·20 g = 0·80 g

contd

DON'T FORGET

The half-life of a radioisotope is the time taken for its mass or activity to halve. The half-lives of some radioisotopes can be found on page 8 of the Data Booklet.

HALF-LIVES contd

Example 2

A sample of radon-222 has an initial activity of 8.0×10^4 counts per second (cps). After 12 days its activity fell to 1.0×10^4 cps. What is the half-life of radon-222?

The activity will halve after each half-life period. So,

$$\text{initial count rate} = 8.0 \times 10^4 \text{ cps}$$
$$\text{count rate after 1 half-life period} = 4.0 \times 10^4 \text{ cps}$$
$$\text{count rate after 2 half-life periods} = 2.0 \times 10^4 \text{ cps}$$
$$\text{count rate after 3 half-life periods} = 1.0 \times 10^4 \text{ cps}$$

This means that in 12 days, three half-life periods have passed and so the half-life of the radioisotope is $\frac{12}{3} = 4$ days.

Example 3

Fluorine-18 is a radioisotope used to diagnose disorders in the brain. It has a half-life of 1·8 hours. How long would it take for the activity of a sample of fluorine-18 to fall to one sixteenth of its original value?

After 1 half-life period, the activity would be half of what it was originally
after 2 half-life periods, the activity would be a quarter of what it was originally
after 3 half-life periods, the activity would be an eighth of what it was originally
after 4 half-life periods, the activity would be a sixteenth of what it was originally.

Four half-life periods have passed for the activity to fall to one sixteenth of its original value and so the time taken for this to happen is $4 \times 1.8 = 7.2$ hours.

DON'T FORGET

You must be able to calculate the quantity of a radioisotope, its half-life or time elapsed given the values of the other two variables.

LET'S THINK ABOUT THIS

Radioactive decay measurements can be used in determining the age of certain things. For example, the radioisotope uranium-238 (U-238) provides a means of dating rocks. It decays by a series of steps to lead-206 (Pb-206) which is stable. By measuring the ratio of Pb-206 atoms to U-238 atoms in the rock sample and knowing that U-238 has a half-life of 4.51×10^9 years, we can estimate the age of the rock. Let's consider an example – analysis of a meteorite showed the Pb-206:U-238 ratio to be 3, that is three Pb-206 atoms for every one U-238 atom. In estimating the age of the meteorite we assume that no Pb-206 was present at the start and that all the Pb-206 atoms now present came from the decay of U-238. We can construct a table showing the number of half-lives of U-238 that have passed and the number and ratio of U-238 and Pb-206 atoms. To make the arithmetic easier, let's say we started with 32 U-238 atoms.

Half-lives of U-238	Number of U-238 atoms	Number of Pb-206 atoms	$\frac{Pb\text{-}206}{U\text{-}238}$ ratio
0	32	0	0
1	16	16	1
2	8	24	3

Since the Pb-206:U-238 ratio in the sample was found to be 3, then two half-lives must have passed since the meteorite's formation. This means that it is $2 \times 4.51 \times 10^9 = 9.02 \times 10^9$ years old.

NUCLEAR CHEMISTRY 3

RADIOISOTOPES

Uses of radioisotopes

Radioisotopes are used in a wide variety of ways and some of these are outlined below.

- **Medical**
Radioisotopes are used in medicine in two distinct ways: diagnosis and therapy. Diagnostic procedures using radioisotopes involve the production of images of specific parts of the body. The radioisotope is administered to the patient either as the element or as a compound and it accumulates in the tissue that is to be imaged. There, the radioisotope decays and the radiation it emits, is detected outside the body and the information collected is converted into a meaningful image. Technetium-99 is the most commonly used diagnostic radioisotope. Incorporated within appropriate compounds, it can be introduced into the body so that it accumulates in various organs including the heart, kidneys, liver and lungs and in glands such as the thyroid. Technetium-99 is particularly suitable in diagnostic medicine since it is a gamma-emitter and has a short half-life of six hours. Gamma rays exit the body much more efficiently than alpha and beta radiations and cause much less damage to the tissues. Technetium-99's short half-life also means that radiation damage is kept to a minimum. Iodine-123 is another radioisotope used in diagnosing disorders particularly those of the thyroid gland. It is introduced into the patient's body as a solution of sodium iodide either as a drink or by injection. Like technetium-99, iodine-123 is useful in diagnostic imaging because it is a gamma-emitter and has a short half-life of 13 hours.

  Although radiation can cause cancer, it can also be used to effect cures. For example, cobalt-60, which is a gamma-emitter, is used in the treatment of deep-seated tumours. The gamma rays from a source outside the body are focused on the tumour and they destroy the cancerous cells. Some healthy cells are also destroyed in the process but they are much less susceptible to radiation damage since they divide less rapidly than cancerous cells. Highly penetrating radiation has to be used here because the rays have to pass through body tissue in order to reach the tumour. When a radioisotope is used externally, then a long half-life period is preferred. It means that the source will remain active for a long period of time and won't require constant replacement. Cobalt-60 has a half-life of five years.

- **Industrial**
Radioisotopes have a large number of industrial applications. For example, strontium-90, a beta-emitter, is used in paper manufacture to monitor and control the thickness of the paper. The set up is illustrated in the diagram below. As the sheet of paper emerges from the rollers it passes between a strontium-90 source and a beta-detector. The amount of radiation reaching the detector depends on the thickness of the paper – the greater the thickness the fewer the number of beta particles detected. If the radiation reaching the detector falls below a certain specified level then this implies that the paper is too thick. This information is fed back to the rollers which automatically move closer together. In this particular application, beta radiation has to be used. Alpha particles would be stopped by the paper and gamma rays would penetrate it completely – as a result, any variations in the thickness of the paper would not be detected. Strontium-90 has a long half-life of 28 years which means that it won't need constant replacement.

DON'T FORGET

Radioisotopes which are ingested must have short half-lives but the half-lives of those used externally should be long.

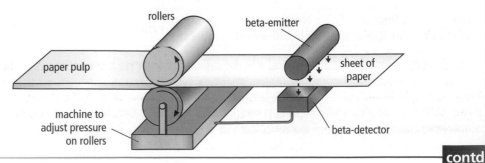

rollers

beta-emitter

paper pulp

sheet of paper

machine to adjust pressure on rollers

beta-detector

contd

RADIOISOTOPES contd

- **Scientific research**

Radioisotopes are used extensively in scientific research. For example, the radioisotope, carbon-14, is used to date archaeological specimens containing material that was once living. It is formed in the upper atmosphere when nitrogen-14 is bombarded by the neutrons present in cosmic rays:

$$^{14}_{7}N + ^{1}_{0}n \longrightarrow ^{14}_{6}C + ^{1}_{1}p$$

Carbon-14 in the form of carbon dioxide is absorbed by plants during photosynthesis and so the tissues of all living plants and animals contain the radioisotope. It decays by beta emission and has a half-life of 5600 years:

$$^{14}_{6}C \longrightarrow ^{14}_{7}N + ^{0}_{-1}e$$

The carbon-14 lost by radioactive decay is constantly replenished by its formation in the atmosphere. In this decay-replenishing process a dynamic equilibrium is established and as a result there is a constant level of carbon-14 in the atmosphere and in all living organisms. However, when the organism dies, it can no longer absorb carbon-14 and so the amount of radiation emitted by that organism will fall. By comparing its activity with that of living material and knowing the half-life of carbon-14, it is possible to work out the age of the specimen. To illustrate the method, let's consider the following example. A wooden utensil recovered from an ancient archaeological site had a count rate of 7·5 counts per minute per gram of carbon. Wood from a freshly cut tree has an activity of 15·0 counts per minute per gram of carbon. This means that one half-life period has passed for the count rate to fall from 15·0 to 7·5 and so the wooden beam must be 5600 years old.

Radioisotopes have also been used to work out the various steps involved in some chemical reactions. For example, the complete mechanism of photosynthesis was deduced using carbon dioxide labelled with the radioisotope carbon-14.

DON'T FORGET

Radiocarbon dating depends on the fact that the rate of decay of carbon-14 is equal to its rate of formation in the upper atmosphere.

LET'S THINK ABOUT THIS

A major application of radiocarbon dating was the determination of the age of the Turin Shroud, believed to be the burial cloth of Jesus Christ. In 1988, three independent radiocarbon analyses were carried out in laboratories in Europe and the United States on samples of the shroud no bigger than a postage stamp. The results showed that it dated from between 1260 and 1390 and is therefore unlikely to have been the burial cloth of Christ.

NUCLEAR CHEMISTRY 4

MORE ON RADIOISOTOPES

Energy production

Nuclear reactions are accompanied by an overall decrease in mass and this mass loss is converted into energy which can then be harnessed.

To produce energy on a large scale, the naturally occurring radioisotope, uranium-235, can be used. When its heavy nuclei are bombarded by neutrons, they undergo **nuclear fission**, that is they each split up into two lighter nuclei, for example

$$^{235}_{92}U + {}^{1}_{0}n \longrightarrow {}^{92}_{36}Kr + {}^{142}_{56}Ba + 2\,{}^{1}_{0}n$$

Notice that one neutron goes in to the fission reaction and two come out. These two neutrons cause the fission of two other uranium-235 nuclei and four neutrons are then generated. As a result, a chain reaction sets in, as illustrated in the diagram opposite. The chain reaction gets faster and faster and enormous amounts of energy are released in a very short time. In a nuclear reactor, however, the chain reaction is not allowed to get out of hand. Its rate is maintained at a reasonable level by the use of control rods made from boron steel. They are lowered into the reactor core and absorb a large proportion of the neutrons and so reduce the number of fission reactions taking place. The energy produced in the fission process is then used to generate electricity.

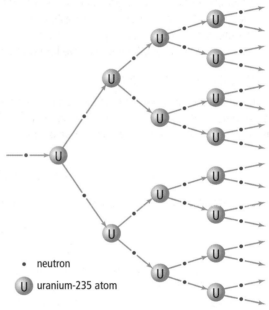

• neutron

Ⓤ uranium-235 atom

DON'T FORGET

In nuclear fusion reactions, light nuclei join together to form heavier nuclei whereas in nuclear fission reactions, heavy nuclei break up into lighter nuclei.

Currently, research is being carried out to look at energy production by way of **nuclear fusion**. In the fusion process, two light nuclei fuse together to form a heavier nucleus, for example

$$^{2}_{1}H + {}^{3}_{1}H \longrightarrow {}^{4}_{2}He + {}^{1}_{0}n$$

The light nuclei have to be heated to very high temperatures so that they collide with sufficient speed to overcome the repulsive forces between them. However, when they do fuse together, the energy released is even greater than that produced in the fission process. There are still major problems to be resolved in the controlled production of energy from fusion but in the future, fusion may well replace fission as the source of nuclear energy.

Nuclear fuels and fossil fuels – a comparison

Energy can be produced from both nuclear and fossil fuels (coal, oil and natural gas) and currently in the UK, about twice as much electricity is generated from fossil fuels than from nuclear fuels. They can be compared in terms of energy efficiency, finite resources, safety and pollution:

- **Energy efficiency**
  For the same mass, nuclear fuels produce much more energy than fossil fuels. For example, 1g of uranium-235 yields about 8 100 000 kJ of energy whereas 1g of good quality coal produces only 30 kJ.

contd

MORE ON RADIOISOTOPES contd

- **Finite resources**
  Both nuclear fuels and fossil fuels are finite. This, however, presents less of a problem with nuclear fuels because of their greater energy efficiency.

- **Safety**
  While the risk of death or injury to workers in the nuclear industry is lower than that in the fossil fuel industry, there are hidden dangers associated with the nuclear industry, for example leukaemia 'clusters' in the vicinity of nuclear power stations. Furthermore, a major accident, like Chernobyl, has much more catastrophic consequences than one in the fossil fuel industry.

- **Pollution**
  When fossil fuels are burned, large quantities of carbon dioxide are released into the atmosphere and being a greenhouse gas, it is a major contributor to global warming. Sulphur dioxide is also produced and it contributes to acid rain. Although the nuclear industry produces neither of these pollutant gases, it is not without its problems. The major one is the disposal of nuclear waste. The waste consists of the lighter nuclei that are formed in the fission process. These fission products are highly radioactive and because many of them have long half-lives, they will remain active for hundreds and even thousands of years.

Elements and the stars

All naturally occurring elements originated in the stars as a result of nuclear fusion reactions. In the centres of stars, temperatures are so high that they allow nuclei to fuse together to form heavier nuclei. For example, helium can be built up by the fusion of two hydrogen nuclei:

$$^{1}_{1}H + ^{2}_{1}H \longrightarrow ^{3}_{2}He$$
$$^{2}_{1}H + ^{3}_{1}H \longrightarrow ^{4}_{2}He + ^{1}_{0}n$$

These fusion reactions continue, building up even heavier elements, for example:

$$3\,^{4}_{2}He \longrightarrow ^{12}_{6}C$$
$$^{12}_{6}C + ^{4}_{2}He \longrightarrow ^{16}_{8}O$$

Layers of elements build up within the star with the heaviest at the core and the lightest on the surface. When the core consists mainly of iron, the star becomes unstable and explodes. The explosion of a star is known as a supernova and it is during this process that the elements heavier than iron are formed. The huge gas clouds resulting from supernovae are made up of a whole range of elements and as they condense they form solar systems including our own.

DON'T FORGET

All naturally occurring elements, including those found in our bodies, originated in stars.

LET'S THINK ABOUT THIS

All nuclear reactions are accompanied by a loss in mass and the energy released (E) is related to this mass loss (m) by Einstein's famous equation:

$$E = mc^2$$

where c is the speed of light ($3 \cdot 00 \times 10^8 \, m\,s^{-1}$).

Let's work out the energy released in the following nuclear fusion reaction:

$$^{2}_{1}H + ^{2}_{1}H \longrightarrow ^{3}_{2}He + ^{1}_{0}n$$

The mass loss in this reaction is $3 \cdot 50 \times 10^{-6} \, kg$. Using Einstein's equation:

$$E = mc^2 = 3 \cdot 50 \times 10^{-6} \times (3 \cdot 00 \times 10^8)^2$$
$$= 3 \cdot 15 \times 10^{11} \, J\,mol^{-1} \text{ or } 3 \cdot 15 \times 10^8 \, kJ\,mol^{-1}$$

To put this in perspective, $3 \cdot 15 \times 10^8 \, kJ\,mol^{-1}$ represents just over a million times the energy released on burning one mole of hydrogen.

PPA 1–3

PPA 1 – HESS'S LAW

DON'T FORGET

Hess's Law states that the enthalpy change for a chemical reaction is independent of the route taken.

Introduction

Solid potassium hydroxide can be converted into potassium chloride solution by two different routes:

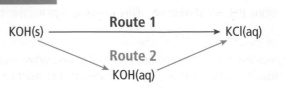

Route 1 is the direct route whereby potassium chloride solution is made by adding solid potassium hydroxide directly to hydrochloric acid. This has enthalpy change, ΔH_1.

$$KOH(s) + HCl(aq) \longrightarrow KCl(aq) + H_2O(l) \qquad \Delta H_1$$

Route 2 is the indirect route and involves two steps. In the first step solid potassium hydroxide is dissolved in water forming potassium hydroxide solution.

$$KOH(s) + aq \longrightarrow KOH(aq) \qquad\qquad \Delta H_{2a}$$

The potassium hydroxide solution is then added to hydrochloric acid to form potassium chloride solution:

$$KOH(aq) + HCl(aq) \longrightarrow KCl(aq) + H_2O(l) \qquad \Delta H_{2b}$$

According to Hess's Law the overall enthalpy change involved in converting solid potassium hydroxide into potassium chloride solution will be the same no matter whether the direct or indirect route is taken, i.e. $\Delta H_1 = \Delta H_{2a} + \Delta H_{2b}$

Aim

The aim of this experiment is to confirm Hess's Law.

Procedure

Route 1 (Direct route)

$$KOH(s) + HCl(aq) \longrightarrow KCl(aq) + H_2O(l) \; \Delta H_1$$

25 cm³ of 1 mol l⁻¹ HCl(aq) was measured out into a plastic beaker and its temperature noted. Approximately 1·2 g of KOH(s) was weighed out accurately into a plastic beaker. The acid was added to the KOH(s) and slowly and continuously stirred until all the KOH(s) had reacted. The highest temperature reached by the reaction mixture was noted.

Route 2 (2 steps)

Step 2a $KOH(s) + aq \longrightarrow KOH(aq)$ $\qquad\qquad\qquad\quad \Delta H_{2a}$
Step 2b $KOH(aq) + HCl(aq) \longrightarrow KCl(aq) + H_2O(l)$ $\qquad \Delta H_{2b}$

Step 2a: 25 cm³ of H₂O was measured into a plastic beaker and its temperature noted. Approximately 1·2 g of KOH(s) was weighed out accurately into a plastic beaker. The water was added to the KOH(s) and it was slowly and continuously stirred until all the solid had dissolved. The highest temperature reached by the reaction mixture was noted.

Step 2b: 25 cm³ of 1 mol l⁻¹ HCl(aq) was measured out into a plastic beaker and its temperature noted. The temperature of the KOH(aq) made in step 2a was noted. The acid was added to the KOH(aq) and after stirring slowly and continuously the highest temperature reached was noted.

Results and calculation

Route 1 – calculation of ΔH_1

Suppose the mass of KOH(s) used was 1·22 g and when added to 25 cm³ of HCl(aq) the temperature rise was 23°C, then the heat energy gained by the reaction mixture,

$E_h = cm\Delta T = 4·18 \times 0·025 \times 23 = 2·4035$ kJ.

Formula mass of KOH = 56, so 1 mole of KOH = 56 g.

contd

PPA 1 – HESS'S LAW contd

The heat energy which would have been released if one mole of KOH (56 g) had been added to hydrochloric acid is then calculated:

$$1.22g \longleftrightarrow 2.299\,kJ$$
$$56g \longleftrightarrow \frac{2.4035 \times 56}{1.22} = 110\,kJ$$

Therefore, $\Delta H_1 = -110\,kJ\,mol^{-1}$. (The negative sign is there because the reaction is exothermic.)

Route 2 – calculation of ΔH_{2a}

Suppose the mass of KOH(s) used was 1.21 g and when added to 25 cm^3 of H_2O the temperature rise was 11°C, then the heat energy gained by the reaction mixture,

$$E_h = cm\Delta T = 4.18 \times 0.025 \times 11 = 1.1495\,kJ.$$

The heat energy which would have been released if one mole of KOH (56 g) had been added to water is then calculated:

$$1.21g \longleftrightarrow 1.1495\,kJ$$
$$56g \longleftrightarrow \frac{1.1495 \times 56}{1.21} = 53.2\,kJ \qquad \text{Therefore, } \Delta H_{2a} = -53.2\,kJ\,mol^{-1}.$$

Route 2 – calculation of ΔH_{2b}

Suppose the temperature rise of the reaction mixture was 6°C when 25 cm^3 of HCl(aq) was added to the 25 cm^3 of KOH(aq) prepared in step 2a, then the heat energy gained by the reaction mixture,

$$E_h = cm\Delta T = 4.18 \times 0.050 \times 6 = 1.254\,kJ.$$

1.21 g of KOH(s) was present in the KOH(aq) so the heat energy which would have been released if one mole of KOH had been added to HCl(aq) can be calculated:

$$1.21g \longleftrightarrow 1.254\,kJ$$
$$56g \longleftrightarrow \frac{1.254 \times 56}{1.21} = 58.0\,kJ \qquad \text{Therefore, } \Delta H_{2b} = -58.0\,kJ\,mol^{-1}.$$

$$\Delta H_{2a} + \Delta H_{2b} = -53.2 + (-58.0) = -111.2\,kJ\,mol^{-1} \qquad \Delta H_1 = -110\,kJ\,mol^{-1}$$

Conclusion

Since the enthalpy change for route 1 (ΔH_1) is approximately equal to the total enthalpy change for route 2 ($\Delta H_{2a} + \Delta H_{2b}$), then Hess's Law has been confirmed.

Evaluation

- It was assumed that all the heat generated in the reactions was transferred to the solutions but some of it will have been transferred to the surrounding air, the reaction container and the thermometer and these were not taken into account.

- Polystyrene cups (or plastic beakers) were used to cut down heat losses since they are poor conductors of heat.

- Heat losses could have been reduced further if lids had been placed on the polystyrene cups or vacuum flasks had been used instead of the polystyrene cups.

- Since both solid potassium hydroxide and potassium hydroxide solution are corrosive and 1 mol l^{-1} hydrochloric acid irritates the eyes, goggles had to be worn and any splashes on the skin had to be washed off immediately.

PPA 2 – REDOX TITRATION

Introduction

Vitamin C can undergo a redox reaction with iodine in which the vitamin C ($C_6H_8O_6$) is oxidised

$$\underset{\text{vitamin C}}{C_6H_8O_6} \longrightarrow C_6H_6O_6 + 2H^+ + 2e^- \qquad \text{oxidation}$$

and iodine (I_2) molecules are reduced to iodide ions (I^-)

$$\underset{\text{iodine molecules}}{I_2} + 2e^- \longrightarrow \underset{\text{iodide ions}}{2I^-} \qquad \text{reduction}$$

$$C_6H_8O_6 + I_2 \longrightarrow C_6H_6O_6 + 2H^+ + 2I^- \quad \text{is the balanced overall redox equation.}$$

> **DON'T FORGET**
>
> In Higher Chemistry, c is always the specific heat capacity of water and its value is given in the Data Booklet. m is the mass of solution which is mainly water.

contd

PPA 2 – REDOX TITRATION contd

Aim

The aim of this experiment is to determine the mass of vitamin C in a tablet by carrying out a redox titration using a solution of iodine of accurately known concentration and starch solution as an indicator.

Procedure

A vitamin C tablet was dissolved in approximately $50\,cm^3$ deionised water and the solution plus all the rinsings were transferred into a $250\,cm^3$ volumetric flask which was then made up to the $250\,cm^3$ mark with more deionised water.

A pipette was used to measure out a $25 \cdot 0\,cm^3$ sample of the vitamin C solution into a conical flask.

A burette was rinsed and filled with iodine solution of known concentration. A few drops of starch solution were added to the conical flask and the titration started.

The end point occurred when the starch indicator just turned a permanent blue-black colour. This indicates that all the vitamin C in the conical flask has been used up and the one drop of excess iodine added has reacted with the starch.

The titration was repeated with more $25 \cdot 0\,cm^3$ samples of the vitamin C solution until concordant results were obtained.

Results and calculation

Typical results from this experiment are shown in the table below:

	Rough	1st trial	2nd trial
Final burette reading (cm³)	23·10	45·30	22·30
Initial burette reading (cm³)	0·10	23·10	0·20
Titre volume (cm³)	23·00	22·20	22·10

The concordant results were averaged and this was the volume of iodine used in the calculation.

Average volume of iodine solution required $= \dfrac{22 \cdot 20 + 22 \cdot 10}{2} = 22 \cdot 15\,cm^3$.

Concentration of iodine solution $= 0 \cdot 0250\,mol\,l^{-1}$.

The number of moles of iodine, $n = V \times c = 0 \cdot 02215 \times 0 \cdot 0250 = 5 \cdot 5375 \times 10^{-4}\,mol$.

The balanced redox equation, 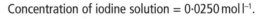 $C_6H_8O_6 + I_2 \longrightarrow C_6H_6O_6 + 2H^+ + 2I^-$
shows that 1 mole of vitamin C reacts with 1 mole of iodine.

So the number of moles of vitamin C in the $25 \cdot 0\,cm^3$ sample is also $5 \cdot 5375 \times 10^{-4}\,mol$.

The total number of moles of vitamin C (in $250\,cm^3$) is $5 \cdot 5375 \times 10^{-3}\,mol$.

The number of moles of vitamin C in the tablet was therefore $5 \cdot 5375 \times 10^{-3}\,mol$.

The formula mass, FM, of vitamin C, $C_6H_8O_6$, is 176.

The mass of vitamin C per tablet $= n \times FM = 5 \cdot 5375 \times 10^{-3} \times 176 = 0 \cdot 97\,g$.

Conclusion

The mass of vitamin C in the tablet was $0 \cdot 97\,g$.

Evaluation

- In preparing the vitamin C solution it was important to:
  1. wash the beaker several times to ensure that all the vitamin C was transferred to the standard flask,
  2. invert the standard flask several times to make sure that the vitamin C solution was thoroughly mixed and had a uniform concentration.
- Prior to carrying out the titration it was important to rinse the pipette with some of the vitamin C solution and the burette with a little of the iodine solution otherwise the solutions could become contaminated.
- In calculating the average titre volume, it is the concordant titres only that must be averaged.
- A white tile was placed under the conical flask – this allowed the colour change at the end-point to be detected more clearly.

DON'T FORGET

Remember, in this type of calculation, the volume in cm^3 must be converted into volume in litres.

DON'T FORGET

To get the total number of moles of vitamin C in the tablet, you have to multiply the number of moles in $25 \cdot 0\,cm^3$ by 10 since the total volume of the vitamin C solution is $250\,cm^3$.

PPA 3 – QUANTITATIVE ELECTROLYSIS

Introduction

Sulphuric acid, like all dilute acids, contains more hydrogen ions than hydroxide ions.

During electrolysis, the positive hydrogen ions travel to the negatively charged electrode where they are reduced to hydrogen gas. The ion-electron equation is:

$$2H^+(aq) + 2e^- \longrightarrow H_2(g)$$

This shows that two moles of electrons are needed to liberate one mole of hydrogen gas.

Since 96 500 C is the charge associated with one mole of electrons, then $2 \times 96\,500$ C will, in theory, be required to produce one mole of hydrogen.

Aim

The aim of the experiment is to confirm that $2 \times 96\,500$ C (193 000 C) is required to produce 1 mole of hydrogen by electrolysing dilute sulphuric acid.

Procedure

The apparatus shown in the diagram was set up but without the measuring cylinder over the negative electrode.

The power supply was switched on and the current set to approximately 0·5 A by adjusting the voltage on the lab pack or by adjusting a variable resistor which could be introduced into the circuit.

The current was passed through the circuit for a few minutes to allow the porous carbon electrodes to become completely saturated with hydrogen gas.

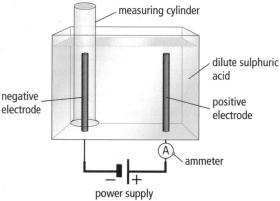

The power supply was then switched off before positioning a measuring cylinder filled with the acid over the negative electrode as shown.

The power supply was then switched on and a timer started. The current was passed through the circuit until slightly less than 50 cm³ of hydrogen gas had been collected. Throughout this time the current was constantly checked and adjusted if necessary. The current used and the time to collect the gas were recorded, as well as the exact volume of hydrogen produced.

Results and calculation

A typical set of results are given opposite:

Conclusion

The quantity of electricity required to produce one mole of hydrogen by electrolysis was 193 238 Coulombs which is very close to the theoretical value of 193 000 C.

> Volume of hydrogen produced = 44·0 cm³ = 0·0440 litres
> Current used = 0·48 A
> Time = 12 minutes 15 seconds = 735 s
>
> Q = I × t = 0·48 × 735 = 352·8 Coulombs
>
> Assuming that under the conditions of temperature and pressure on that day, the molar volume of hydrogen is 24·10 litres mol⁻¹.
>
> Then 0·0440 litres was produced by 352·8 C and so
>
> 24·10 litres would be produced by $352 \cdot 8 \times \dfrac{24 \cdot 10}{0 \cdot 0440} = 193\,238$ Coulombs.

Evaluation

- The sources of error which could account for the difference between the experimental result and the theoretical one are:
    1. the current may not have remained constant during the electrolysis – a variable resistor may have been introduced into the circuit to minimise this effect
    2. some of the hydrogen may have dissolved in the sulphuric acid – this would have led to an underestimate of the volume of hydrogen being measured
    3. the molar volume of the hydrogen may not have been 24·10 litres mol⁻¹ under the conditions of temperature and pressure of the experiment.

- Before the measuring cylinder was placed over the negative electrode, it was important that the dilute sulphuric acid was electrolysed for a few minutes in order that the porous carbon electrode became saturated with the hydrogen gas. Were this not carried out, the volume of hydrogen produced would have been underestimated. The use of platinum electrodes rather than carbon electrodes would have eliminated this effect.

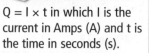

DON'T FORGET

Q = I × t in which I is the current in Amps (A) and t is the time in seconds (s).

LET'S THINK ABOUT THIS ANSWERS

Page 7
(i) $0.67\,cm^3\,s^{-1}$
(ii) $0.25\,cm^3\,s^{-1}$
(iii) $0.12\,cm^3\,s^{-1}$
(iv) $0.033\,cm^3\,s^{-1}$

Page 9
1 The reaction is exothermic and the heat given out provides the remaining reactant particles with the energy they require to collide successfully.
2 Reactions such as neutralisation which take place at room temperature have low activation energies.

Page 11
A catalyst lowers the activation energy for the reaction and so more reacting particles have energy greater than or equal to the lowered activation energy. Increasing the temperature gives the reacting particles more energy and so more reacting particles have energy equal to or greater than the activation energy. This time the activation energy has not been lowered.

Page 13
1 The particles may not collide with enough energy – they must hit each other with energy equal to or greater than the activation energy.
2 The activated complex may break back down into the reactants rather than changing into the products – it is the same activated complex for the reverse and forward reactions.

Page 15
1 The catalyst and the reactants are all in the same state – aqueous solution – so it is a homogeneous catalyst. The pink $Co^{2+}(aq)$ ions change to green $Co^{3+}(aq)$ ions when they catalyse the reaction but revert back to pink $Co^{2+}(aq)$ ions when the reaction is over.
2 To answer this question it is best to draw the potential energy diagram which shows both the catalysed and uncatalysed reactions. The answers are (i) $215\,kJ\,mol^{-1}$ (ii) $113\,kJ\,mol^{-1}$ (iii) $65\,kJ\,mol^{-1}$.

Page 19
(i) $24.7\,g$
(ii) $0.48\,g$

Page 21
(i) Electronegativity is a measure of the attraction an atom has for the electrons in a bond and the noble gases do not form bonds so do not have electronegativity values.
(ii) Lithium has only three electrons.
(iii) Aluminium atoms have three electrons in their outer shell. The fourth electron to be removed is from the second shell which is much closer to the positive nucleus and so much more strongly attracted. Removal of this fourth electron requires much more energy.

Page 23
The electronegativity difference is the same (1·3) each time. Water has polar covalent bonding and is liquid at room temperature. Sodium hydride is ionic and so is solid at room temperature and conducts when molten. It has ionic formula Na^+H^-. This should show that electronegativity values may be helpful in predicting the type of bonding but definite evidence comes only from a study of the properties.

Page 25
1 C_2H_5OH, CH_3NH_2 and HF since they contain hydrogen atoms bonded to oxygen, nitrogen and fluorine, respectively.
2 HCl, CH_3Cl, $CHCl_3$, H_2S and NH_3 have polar molecules.

Page 27
Silicon dioxide has a covalent network structure and so covalent bonds are broken at its melting point.
Sulphur, phosphorus and phosphorus hydride have non-polar covalent structures and so van der Waals' forces are broken at their melting points. Zn is a metal and so metallic bonds have to be broken. Hydrogen bonds between water molecules and ammonia (NH_3) molecules would have to be broken. Ionic bonds would have to be broken in sodium oxide and aluminium oxide.

Page 29
1 (a) 1.505×10^{23} (b) 1.88×10^{23} (c) 1.18×10^{23}
2 (a) 2.06×10^{22} (b) 1.204×10^{22} (c) 1.37×10^{22}
3 (a) 3.01×10^{24} (b) 5.31×10^{23} (c) 4.90×10^{23} (d) 2.01×10^{23}

Page 31
The balanced chemical equation is

	$C_2H_4(g)$	+	$3O_2(g)$	$\rightarrow$	$2CO_2(g)$	+	$2H_2O(l)$
The mole ratio is	1 mol		3 mol		2 mol		2 mol

The ratio of
reacting volumes is   1 volume   3 volumes   2 volumes
Therefore $200\,cm^3$ $600\,cm^3$ $400\,cm^3$
(a) The volume of excess oxygen = $900 - 600 = 300\,cm^3$
(b) The gas at the end of the experiment = $400\,cm^3$ of CO_2 formed plus $300\,cm^3$ of unreacted oxygen giving a total volume of $700\,cm^3$.

Page 85
1 When pH = 4, $[H^+] = 1 \times 10^{-4}\,mol\,l^{-1}$ and when pH = 6, $[H^+] = 1 \times 10^{-6}\,mol\,l^{-1}$. Therefore the concentration of the H^+ ions have decreased by a factor of 100 (Answer D).
2 pH = 2·5 and so the pH is between 2 and 3. The $[H^+]$ is between 1×10^{-2} and $1 \times 10^{-3}\,mol\,l^{-1}$ and therefore answer C is correct (between 0·01 and 0·001 $mol\,l^{-1}$).
3 In the lemon juice the pH = 3, and so $[H^+] = 1 \times 10^{-3}\,mol\,l^{-1}$. In the apple juice the pH = 5 and so $[H^+] = 1 \times 10^{-5}\,mol\,l^{-1}$.

Therefore the ratio is $\dfrac{10^{-3}}{10^{-5}} = 100:1$ (Answer A).

Page 87
1 Answer D, stoichiometry of reaction is same for both strong and weak bases.
2 Answer C, ammonia is a weak base and so is not completely ionised but, like all alkaline solutions, will contain more hydroxide ions than hydrogen ions

INDEX